WAGING BUSINESS WARFARE

Also by David J. Rogers

FIGHTING TO WIN
AGENCY AND COMPANY

WAGING BUSINESS WARFARE

*Lessons from the Military Masters
in Achieving Corporate Superiority*

David J. Rogers

CHARLES SCRIBNER'S SONS
New York

Library of Congress Cataloging-in-Publication Data

Rogers, David J.
Waging business warfare.

Includes index.
1. Management.   2. Success in business.
3. Military art and science.   I. Title.
HD38.R5776   1987   658.4   86-31584
ISBN 0-684-18596-2

Published simultaneously in Canada by Collier Macmillan Canada, Inc.

Composition by Maryland Linotype, Baltimore, Maryland

Manufactured by R. R. Donnelley & Sons, Harrisonburg, Virginia

Designed by Marek Antoniak

First Edition

This book is dedicated to my parents, Bernice and Idwal Rogers. They taught their children to love ideas and to hear the music in the written word.

ACKNOWLEDGMENTS

I wish to thank John A. Rogers for his assistance in this project. How fortunate I have been to have known all my life this student of the military arts—my brother John.

My secretary, Nancy Goggin, helped immeasurably in organizing the research for *Waging Business Warfare* and typing the numerous drafts that eventually became a book. Thanks, Nancy.

Editors never receive enough credit. Mine was Susanne Kirk, executive editor of Charles Scribner's Sons. Her enthusiasm for the book never faltered.

I am grateful to the executives, managers, and staff who answered my questions and shared their views and experiences with me.

A special acknowledgment and thank-you goes to Diana, Stephanie, Alice, Evan, and Eli for their aid in compiling research, preparing drafts, and showing continued support.

CONTENTS

WAGING BUSINESS WARFARE

The Art of Business Warfare

A Prince should therefore have no other aim or thought,
nor take up any other thing for his study, but war.
Machiavelli

If you work in the business world, you have already shared many experiences with the military masters. For example:

Decentralized management. British general Sir William Slim wrote, "Nothing is easier in jungle or dispersed fighting than for a man to shirk. If he has no stomach for advancing, all he has to do is to flop into the undergrowth . . . join up later, and swear he was the last to leave." Now any mortal who has ever worked has at one time or another ducked out of the office, store, or plant to flop into one kind of undergrowth or another—an afternoon at the golf course, for example. And certainly every corporate manager of a decentralized operation has paused periodically to wonder just how much flopping is going on out there in the field.

Consultants. In ancient China, scholars moved from one province to another offering guidance to sovereigns. But these early consultants were probably more careful with their advice than their counterparts of today. For although they were richly rewarded if their recommendations were good, if their advice turned out not to work, they were sawn in two, boiled, or tied between two chariots and ripped in half.

3

Job rotation. American managers had to relearn from so-called Japanese management what the Japanese had learned from us—the value of orienting employees to the corporate culture through a number of short-term job and departmental assignments. While we might wish to take credit for the concept of job rotation, in fact it was used by the Roman legions. Every Roman soldier who was promoted was rotated from the first cohort through the tenth and back through all the others at a higher and higher rank. Thus, when a man assumed leadership as a centurion, he had personally commanded every cohort and was known by and knew every man. Roman soldiers were also required to put aside half their pay "so that they might not squander it," an ancient payroll-savings plan.

Criticism. Any working person who has ever been singled out and called on the carpet and somehow survived the ordeal knows precisely what Winston Churchill felt when he wrote, "Nothing is more exhilarating than to be shot at without result."

Interunit conflict. So prevalent in corporate life as to require no comment from me at all is the bitter, frustrating truth of war and business that "two branches of a staff can get more hostile to each other than to the enemy."

Apparel. What else but an early version of the dress code of IBM and financial institutions and the dress-for-success school of interpersonal relations is the observation of an eighteenth-century Frenchman: "A well dressed soldier has more respect for himself. He also appears more redoubtable to the enemy and dominates him; for a good appearance is itself a force."

Training and development. Is there a corporate specialist in human-resource development who could argue with the wisdom of not asking employees to perform jobs that they have not first been trained to handle? Or, ". . . the soldier, high or low, whatever rank he has, should not have to encounter in War those things which, when seen for the first time, set him in astonishment and perplexity; if he has only met with them one single time before, even by that he is half acquainted with them."

Productivity. Personal experience and studies of employee productivity alike bear witness to Sun Tzu's twenty-four-hundred-year-old observation: "During the early morning spirits are keen, during the day they flag, and in the evening thoughts turn toward home."

Strategy development. Even the most up to date and rigorous research on corporate decision making confirms the fact that strategy that is purportedly formulated for tomorrow is usually retrograde, rooted in yesterday and one war behind. "The war the Generals always get ready for is the previous one," wrote Major Henry Tomlinson.

The experiences of war are in many ways similar to our own experiences in business. But the history of war and the insights and actions of great military figures offer more than these random similarities. They hold out to us clear-cut lessons on how to manage people and how to achieve competitive superiority: the Biblical story of Gideon, for example. If you're going to be a warrior, you might as well have the name for it. Gideon did—"one who cuts down."

Gideon's leadership task was to "cut down" the Midianites, who were encamped in a valley on the Plain of Esdraelon and great in number, "like grasshoppers." For seven years these powerful tyrants had terrorized Israel, raiding its lands, rustling its livestock, and in general impoverishing Gideon's people.

In a bold act of defiance Gideon toppled the altar of the Midianite god Baal, leveled the sacred grove surrounding it, and then sent messengers throughout the land to gather an army together. Thirty-two thousand men reported. Deciding that this was far too many for one man to handle, Gideon implemented a selection test of mettle. He suggested that anyone who was at all afraid should return home. Twenty-two thousand accepted the invitation, leaving ten thousand men.

Then Gideon applied another selection test, this one of the candidates' savvy. He had all ten thousand go down to the nearby river and drink. The river was dangerously close to the Midianites, and to drink without keeping your eyes in the direction of their camp was foolhardy. Experienced, sharp-witted men would certainly take that simple precautionary measure. Nine thousand seven hundred failed the test by kneeling down and drinking the water. Only three hundred kept their heads up and brought the water up to their mouths in their cupped hands. These were the three hundred candidates Gideon selected.

Wisely, Gideon chose a night action. Under the cover of darkness he crept within earshot of the enemy's camp, close enough to overhear a Midianite sentry describe a dream he'd had to another guard. In the dream, a cake of barley fell into a Midianite tent and overturned it, leaving it flat on the ground. The other sentry interpreted the dream as meaning that the cake of barley—the sword of Gideon— would destroy the camp. Encouraged by this sign of low morale among the enemy, Gideon returned to his men and called them together.

He assigned them to one of three companies and equipped each man with a trumpet, an empty pitcher, and a lamp within his pitcher. "Keep your eyes on me and my company," he told them, "and do what we do." Then he gave them the rallying cry, "The sword of the Lord and of Gideon."

Around midnight the three companies went in close to the camp, and Gideon gave the signal. The three hundred blew their trumpets, broke the pitchers to expose the lighted lamp inside, and shouted the rallying cry. Havoc broke loose in the Midianite camp. Believing they were under attack by a large force and unable to see in the darkness, the terrified Midianites fought each other, stumbled about awkwardly, and fled in every direction.

Gideon then quickly sent messengers to the tribes of Israel, ordering the pursuit and annihilation of the disorganized rabble. The Midianites were cut down in short order.

Although only briefly described in the Book of Judges, Gideon's victory has provided guidance to many eminent commanders. The story is also rife with lessons for corporate managers of today, among them these:

- The leader's personality, self-confidence, and clarity of purpose can go a long way toward winning the hearts and minds of subordinates.

- The leader can create a legend about himself (or herself, as the case may be) through dramatic acts that his followers secretly desire, but lack the guts or imagination, to perform themselves.

- The leader must know his own people—their technical abilities and their strength of character, as well.

- The fruits of preplanning are many; attention to the minutiae, the small details, pays dividends.

- Fight the battle that best suits your resources. If you can't overwhelm your competitors, outwit them. The result is the same.

- Size is a much exaggerated factor. A handful of good troops is better than a host of disorganized ones.

- Pay close attention to morale, the will to win—your own troop's and the competition's. Inspire your own people's and depress the competitor's. Your competitor is already weakened if he fears you.

- Equip your forces properly and at the right time for the job at hand. If everyone needs a trumpet, make certain they have one.

- Reconnoiter personally, especially before critical events. Get out of the office and see for yourself.

- Surprise upsets the competition. People fear what they can't see or comprehend.

- Keep your personnel fully informed of your plans and let them know you will be in the action right along with them. Give them a person and an idea to rally around.

- When you have the competition on the run, press the pursuit to the limit. Go all out.

- Larger units should be broken into smaller ones and coordinated well.

- Keep plans beautifully simple. Complicated plans confuse; simple ones can be followed.

THE LESSONS OF THE MILITARY MASTERS

Do not attack the enemy when he adheres to the rules of war, but profit from his slightest mistakes without delay.

Frederick the Great

Napoleon's methods were a total mystery to many of his opponents. They considered his army's endless mass marches hither and thither across the face of Europe to be totally without logic or rational planning. There seemed to be no understandable, systematic pattern to the man's moves. Antoine Henri Jomini (1779–1869), chief of staff of one of Napoleon's marshals, was first to argue that the contrary was true. According to Jomini, if you looked closely at how Napoleon waged his campaigns and battles, you would discover a coherent concept of warring, a pattern that was anything but mysterious or irrational. Napoleon came across a book written by Jomini. Impressed, he had the author report to him. Napoleon was the kind of leader who mulled over the details of his plans for a long time before finally committing them to paper. When he met with Jomini that day in 1806, his mind was working on the Jena campaign. When the meeting ended, Jomini asked the emperor's permission to meet with him again in four days in the city of Bamberg.

Napoleon was shocked. Bamberg was his destination, and he fully intended to arrive there four days later, but no one was supposed to know.

"Who told you I'm going to Bamberg?"

"The map of Germany, sire," replied Jomini, "and your campaigns of Marengo and Ulm."

Jomini demonstrated that all of Napoleon's successful actions were based on the highly skillful application of certain principles of strategy and tactics. They hadn't been invented by Napoleon but had been valid—and would continue to be—throughout history. Wrote Jomini, "There have existed in all times, fundamental principles on which depend good results in warfare. . . . These principles are unchanging, independent of the kind of weapons, of historical time and of place." He added that if a leader disregards this small number of principles, he places himself in jeopardy, whereas history shows that whenever they had been applied, the leader has "been crowned in nearly every case with success."

Subsequently, Napoleon himself made it clear that he was quite aware of the existence of these time-tested principles of war and of their immense importance. It was because they stuck to the principles that the great military masters such as Hannibal, Alexander, Caesar, Scipio, Gustavus Adolphus, and Frederick the Great had achieved their victories. He wrote, "All great captains have done their great deeds by conforming to the rules and natural principles of their art. . . . They have only succeeded by conforming to rules, whatever might have been the boldness of their designs. . . ." Like virtually all masters of war, Napoleon avidly studied his predecessors' books and their particular style of waging war. Many victories have been won by a well-read general, with a sharp memory, using strategic or tactical moves that were employed hundreds, even thousands, of years before. "Historical examples clarify everything," wrote Clausewitz. "This is particularly true of the art of war."

It is now widely, almost indisputably, accepted that there are a small number of principles of success in war and that they have proved universally valid throughout time and in whatever part of the world war has been waged. No matter when they were written or which master wrote them, treatises on war reveal a stunning agreement on the winning principles. These principles have brought victory in the past; they will in the future.

Anthropologist Harry Holbert Turney-High believed that the existence of these principles raised warfare to a science. War, he observed, is a social science, because it depends on teamwork. It was the first social science to become truly scientific, "for it is the first the practice of which has been reduced to a few simple principles which are true without regard to time or place." Turney-High found war's status as the first scientific social science so obvious that he thought it strange that no one before him had made this observation.

The Lessons of Business Warfare

It's generally agreed that there are eight principles of war. Traditionally, each is condensed into one word or a few words to form the following, a kind of checklist of principles:

Objective

Concentration (also called the principle of mass)

Offensive Action

Mobility (also referred to as movement or maneuver)

Surprise

Security

Economy of Force

Cooperation (sometimes called unity of command)

In *Waging Business Warfare* I have chosen to avoid the traditional way of expressing the principles in one-, two-, or three-word designations. They are too abstract in that form. For example, the single word "concentration" doesn't communicate much of a lesson. "Concentrating Greater Strength at the Decisive Point," the title of chapter 4, conveys the essence of the concentration lesson more accurately, and words a few pages into chapter 4 communicate more—bring together greater strength than the competitor at a single decisive point whose possession promises competitive advantages. "Objective" is less meaningful for the busy business person than chapter 3's title, "Maintaining Your Objective; Adjusting Your Plan."

Each chapter of this book communicates a lesson—actually, a number of lessons—applicable to the business person operating in today's corporate world. You are shown what the best military minds have concluded about each principle and how they have used them in action. Throughout, the insights of the geniuses of war are applied to corporations of today. Scores of examples of companies of all sizes and a variety of industries are used. They demonstrate that the same principles and approaches that brought success to Hannibal, Frederick the Great, and others in warfare bring equally successful results when applied by managers of corporations.

In many instances it's clear that corporations of today are just now learning what the military masters have long known. For example, "microspecialization" will be the retailer's strategic buzzword of the late 1980s and the 1990s. It will be grounded on targeting of merchandise selection, advertising, and store design on clearly differentiated sets of consumers. It is a concept that, if executed well, will bring Sears and other retailers competitive victories, but it is by no means a new concept. Alexander the Great never ran a chain of stores, but he showed more than two thousand years ago that a fine, hairbreadth concentration of power wins battles.

Studies of business enterprises are loudly proclaiming the need to define "corporate cultures" clearly. Twenty-three hundred years ago a Spartan general wrote that the safest course for a great army was to be animated by one spirit; and while many corporations are beginning to learn that small, semiautonomous groups can increase productivity, the Romans had well documented the fact a few thousand years ago. That the military masters came upon insights into competitive superiority before the art of management did shouldn't be surprising. The study of business management as we know it really only began in the twentieth century, while masters of war started to give serious thought to organizing and managing about the time Pharaoh Thutmose III devised a special formation to beat the Syrians in 1439 B.C.

Bobby Knight, the coach of the Indiana University basketball team, reads Sun Tzu's *Art of War* before every season. Pepsi, possibly the most aggressive participant in all of business warfare, wishes to make samurai marketers of its executives and teaches them karate. The company spent eight million dollars just to build a gym for training.

Borrowing the best wisdom from another field and applying it to business is certainly not unknown. Corporate managers are usually

more open-minded than most people, because their success depends
on being receptive to anything that helps them improve. For example,
"systems" approaches that have so improved corporate operations
were taken from the place they were first developed—the field of
biology. The Pepsi and Bobby Knight examples show that military
wisdom is being applied to pursuits other than war. What these
pursuits have in common is competition. The principles and lessons
of warfare are more universal than has been realized. They are
principles and lessons of strategy, tactics, and leadership that apply
equally well to intense competition of any kind, including the corporate competition for markets and profits.

For example, as chapter 4 points out, armies win if their leaders
are savvy enough to select decisive points and to concentrate forces
there that are greater than or better than the opponent's. Of the
seventeen hundred new corporations started every day in the United
States, those that will succeed will do so by concentrating in just the
same way. Chapter 5 points out that in business warfare, as in "pure"
or military warfare, victory can't be achieved unless you take the
offensive at some point. Some have suggested that this doesn't apply
to very large corporations and that they can rest on the defense—
and their laurels and past successes. That's a myth, pure and simple.
Defense is a device for forestalling defeat; it is never a tool of competitive superiority. Almost without exception, whenever huge Procter
& Gamble is beaten, in the United States or more recently in Japan,
it is because it has failed to adhere to the single lesson of "Taking
the Offensive and Maintaining Mobility."

The above are just a few examples of the applicability of lessons
of war to corporate competition. *Waging Business Warfare* points
out many others. Differences between military warfare and the competition between corporations are obvious. Casualties in war are
people. In business warfare they are sales and profits, in some cases,
jobs or careers. In virtually every other respect what holds true for
generalship is equally true for enlightened management of corporations of any size—small, medium, or large.

Competitions Within the Competition

Throughout this book the existence of "competitions within the competition" of business warfare are pointed out. Here is what they refer to:

It's obviously correct to say that business involves competition—GM's competition with Chrysler, Ford, and other automakers; Beech-Nut's with Gerber in the baby-food industry; and IBM's with Digital, its most threatening competitor, and with other major computer companies as well as with the elusive, fast-moving "cloners," etc. It's correct but incomplete. The notion of "competitions within the competition" claims that the competitive struggle between corporations in fact involves many competitions, a number of separate contests along the lines of *each* of the principles of war—the principles of intense competition. At the weary end of a long Sunday afternoon of playing NFL football, one team can say, "Our team won." That's the major competition. But in fact a number of competitions were won by the victor. Its game plan was superior to the losers', for example, and/or its offense was better, or the defense or special teams dominated, or one coach or quarterback was better than the other at calling the right plays.

The principles of business warfare enable a corporation to break competitive victory or loss into components. Each principle, each lesson, is one component of corporate superiority. An overall victory in business warfare is the result of a number of separate victories—your leadership is better than theirs, or your firm concentrates at key points better, or your offensive capabilities are superior, etc. The idea of "competitions within the competition" provides you with a finite number of dimensions for identifying where your firm is particularly strong and where things should be improved.

A Summary of Lessons

Nothing is as certain in business and warfare but that one side won't win. It's a tenet of all of war—including corporate wars over market share, sales, return on investment, and profits—not to assume that the rival, the competitor, won't appear, but knowing that he will, to be ready to meet him; not to presume he won't attack but to expect him to and to make yourself invulnerable no matter what he does. Throughout history, military masters have all wrestled with one fundamental question, the same question that preoccupies today's conscientious corporate managers: "How can we win?" Neither war nor business is a game of chance. There are correct answers to "How can we win?" Particularly competitive companies follow the same principles of war that Hannibal followed. The masters' answers are contained in the lessons of *Waging Business Warfare*. They can be summarized as follows:

1. *Leading.* No competition within the competition of business warfare is more important than the contest pitting the quality of one company's leaders against that of its competitors. Good leadership is the first prerequisite of competitive superiority. "There are no weak soldiers under a strong general," goes the military maxim. Leadership is not specifically a principle of war but a factor permeating each principle. Business warfare is waged between people. Every strategic or tactical move made by any company of any size whatsoever—from the biggest, GM, to the smallest computer-supply store—is a reflection of the minds and personalities of the company's leaders.

   Recently, an economist stated that American corporate people have lost "the habit of competing." The geniuses of business warfare haven't. "Competitiveness means taking action," says GE's chairman John F. Welch. This kind of willingness to take action, to compete, is one of the basic characteristics of the person who possesses a genius for business warfare. Decisive style of leadership is another. William Smithburg, chairman of Quaker Oats, brought the notion of planning aggressively and taking chances to the company—a boldness even in the face of uncertainty that sets the genius of business warfare apart from less successful leaders. But boldness should stop at the limits of probability. In the best leaders it does; prudence also marks exceptional leadership. "Character"—knowing what you want and having the determination and courage to get it—is another element of the person with a special aptitude for business warfare. Other characteristics of the genius of business warfare are being great at critical moments, like Lee Iacocca; showing grace under pressure, like Roger Smith, chairman of GM; strong willpower; innovativeness; concern for company morale; and a strategic sense, like building magnate Donald Trump's "to see what's going to happen."

2. *Maintaining your objective; adjusting your plan.* Winners see only one thing—the main objective. Comparative studies of successful and similar but unsuccessful corporations show that every high-performing firm follows what is in fact the principle of war called the maintenance of the objective. They pursue one or a few great decisive aims with full force and determination. Bell & Howell managers speak of their company's "soul"— the purposes or "superobjectives" that animate its work force.

At the same time that competitively superior companies don't budge from the relentless pursuit of their main objectives, they do budge from their plans—if those plans prove unrealistic. A Booz Allen survey of three thousand senior managers revealed that almost 90 percent felt that their current planning techniques were inadequate. Warfare's "No plan survives contact with the enemy" becomes business's "No plan survives even the first contact with the realities of the marketplace." All other factors being more or less equal, the corporation that can adapt and maneuver better than its competitors will win. TRW, the six-billion-dollar conglomerate, provides a case of a major corporation that won by maintaining its objective when it ventured into the credit-data business but adapted to reality by changing its plan. GE learned the hard way to adapt and change; KitchenAid didn't, and because of it, KitchenAid lost about half of its market share in just a few years' time. TRW also learned the value of the simple axiom of war "Avoid a pitched battle when your opponent is stronger," more memorably expressed as "Never get into a pissing contest with a porcupine."

3. *Concentrating greater strength at the decisive point.* Competitively active corporations, like successful armies, base their strategy and tactics on some form of concentration. They focus more strength than competitors at points that give the corporation the advantage. Beech-Nut in the U.S. seven hundred-million-dollar-a-year baby-food business, Hershey in our over eight-billion-dollar candy market, State Farm in insurance, *Inc.* magazine, Colgate-Palmolive, and Gulf & Western are a few examples of corporations that are profiting from concentrating. The war principle of concentration is generally considered the most important of all. That the side that concentrates best wins is proved every day in business warfare. Napoleon was the master of concentration of force in war; Alfred P. Sloan of GM was perhaps the greatest concentrator in business warfare.

The winning corporations in this competition within the competition select decisive points and concentrate superior force there. The losers either don't concentrate, concentrate at less profitable points, or do so at the right points but with inferior force. Winners compete in ways that maximize their strengths against the competitors' weaknesses in the most opportune markets and with the right products. Segmenting markets, finding your niche, and differentiating your services or products are examples of concentration in business warfare. In business warfare there are two different styles of fighting along lines of concentration—from your core strength or against your competitors' vulnerabilities. IBM represents the first style. Tambrands, Kimberly-Clark, Nabisco, Keebler, and Pepsi-owned Frito-Lay concentrate many of their competitive moves against their opponents' vulnerabilities. It can be shown that concentration wins in industry after industry, yet concentration is one of the most often violated principles. Why? Because to concentrate on certain opportunities means taking a chance on not concentrating on others. That takes a clear corporate head and a great deal of courage.

4. *Taking the offensive and maintaining mobility.* Chapter 5 of this book discusses how two closely connected principles of war apply to business—the principle of offensive action and the principle of mobility. The former states that while it's true that your company may hold off defeat by staying on the defensive, you cannot win through defense. Only by taking the initiative and making your competitors dance to your tune can you win in business warfare. Motorola's position improved sharply shortly after it took to the offensive against Japanese competitors— "tweaking their noses" in electronic components, radio equipment, data communications and computers, as the company chairman put it.

Your company cannot achieve victory unless at some point you take the offensive—to reach a new market, introduce new products or innovations or pricing strategy, or to seize competitors' accounts, etc. You needn't be the first to the market with a new product to eventually gain the biggest share, as Coke, Pepsi, IBM, Digital, and Hewlett-Packard demonstrate. But to muscle your way over the competition, the time must come when you shift to the offense. Mobility is moving and maneuvering decisively and rapidly. In business warfare competitive power is almost totally useless unless you can bring it into action quickly.

IBM dethroned Apple in personal computers by using an

attack maneuver of "penetrating the center," but Gillette success-fully used the maneuver of "enveloping one flank" and found a winner in its Brush Plus shaving-cream dispenser. Those offen-sive moves and the four other basic maneuvers of warfare are being applied today by firms as disparate as accountants and lighter manufacturers. The best of them reveal two other char-acteristics of effective competing—their offensives consist of a series of smaller interlocked offensives, and they pursue, pour-ing more into a strategy that's proving successful.

5. *Following the course of least resistance.* Throughout history most military masters have risked almost anything to avoid facing an enemy that was ready and waiting. Consistently, they have found the safest and most promising course to be the course of the opponent's least resistance. General Foods (GF) was forced to withdraw a line of baby food because it couldn't beat Gerber's matching line. On the other hand, discount brokerages like Schwab & Company succeeded by offering no-frills services. They don't employ commissioned salespeople, and they don't give advice. No frills. GF didn't aim at the competitor's area of least resistance; Schwab did. GF lost; Schwab won. Corporate marketing failures often result from the failure to follow this course but to try instead to attack a well-entrenched competitor. The lessons in chapter 6 include those taught by the war prin-ciple of surprise but are far more comprehensive than surprise alone.

No style of warfare follows the course of the opponent's least resistance as fully as guerrilla warfare. Guerrilla fighters and corporations competing guerrilla style base their entire strategy and tactics on attacking areas of low resistance. Skillfully managed guerrilla corporations, no matter how small, are ex-tremely hard to beat. As one military master said, "The ad-vantages are nearly all on the side of the guerrilla" because he is bound by no rules. Wal-Mart began as a guerrilla operation in Newport, Arkansas, and currently employs more than ninety thousand people. Its founder, Sam Walton, is now the richest man in the United States. McDonald's fought guerrilla style, and many of the nation's up-and-coming corporations are guerrillas.

Wherever you find them, these guerrilla companies, if they are to become large, will pass through the five predictable stages of guerrilla warfare.

6. *Achieving security.* If your competitor suddenly becomes very active and you notice a great deal of movement over there, you can expect a competitive move. If an advantage is clearly open to your competitor and he doesn't take advantage of it, you can assume he's having some problems. These are "signs." The principle of security says recognize them and interpret them correctly, then use what you know to your advantage.

One competition within the competition of business warfare is for intelligence—accurate and useful information on your competitors, their leaders, and your consumers. "Success in war is obtained by anticipating the plans of the enemy and by diverting his attention from your own designs." Analyzing your competitors' strengths and weaknesses and judging their leaders' personalities and reactions are basic parts of a secure company. You also need consumer information. The company that doesn't know its consumers is never secure. Procter & Gamble, Frito-Lay, and the two former guerrillas, McDonald's and Wal-Mart, realize what great generals have also understood: you should never be satisfied with the descriptions of the terrain in which you're operating but should reconnoiter for yourself.

One of the most potent of all the lessons about achieving security is that the best defense is an aggressive offense. The business that entrenches its managers and keeps them cowering in their offices is toying with defeat. Even Royal Crown—a small company compared with competitors Coca-Cola and

PepsiCo—demonstrates that the transition from defense to offense is a necessary stage of competing.

7. *Making certain all personnel play their part.* Martin S. Davis, chairman of Gulf & Western, said, "Bigness is not a sign of strength. In fact, just the opposite is true." And the military master eighteenth-century Frenchman Maurice de Saxe proclaimed, "It is not big armies that win battles; it is the good ones."

The fundamental lesson of the war principle of economy of force has a lot of common sense to it. A corporation (or small unit for that matter) should be designed so that the greatest possible impact of the whole organization can be quickly and simultaneously concentrated on the main issue—the major objective to be reached or the problem to be solved. Davis and de Saxe were right. Bigness alone doesn't account for much. If one corporation employs one hundred thousand people and another has ten thousand working for it, but the smaller has five thousand of its people manufacturing and selling a product and the big company has only two thousand assigned to its competing product, the effect will be the same *as if* the small company were more than twice as big as the larger. Napoleon was usually outnumbered in total, but by skillfully shifting his troops, he actually brought more men into battle than his opponents. Corning Glass, Monsanto, CNA Insurance, and IBM are a few examples of corporations that are actively striving to carry out what Napoleon called "the great art of war"—the art of proportioning manpower to the objective to be achieved and the obstacles to be overcome.

Making certain that all personnel play their part also involves the people-managing aspect of management. Companies, like armies, will go only as far as their personnel will take them and not one inch farther. Some people will always find a reason for staying put and not budging. Good troops and good employees always want to get into the action. They possess the constellation of subjective human qualities called "the fire and movement habit" in *Waging Business Warfare.* Eastman Kodak, Ford, Bell & Howell, and Shenandoah Life Insurance are finding ways of increasing that habit by creating teams of firers and movers —fire teams.

YOU'RE THE IMPORTANT ELEMENT

*. . . the combination of wise theory with great
character . . . will make the great captain.*

Antoine H. Jomini

Jim and I went to meet with managers of one of my consulting clients. Jim was very bright, extremely well educated, with a Ph.D. in organizational psychology, and had studied with some of the world's top experts on morale and productivity. During the meeting the client's people discussed their problems, and I was feeling terrific —not because the client had problems but because I knew that with Jim's background he'd be able to work out solutions for them. I kept thinking, *This is right up Jim's alley.* Wrong again.

After the meeting, back at our office, I sat down with him, gulped down a slug of lukewarm coffee, and said excitedly, "Well, what do you suggest we do for them?" *He didn't know.* He could state the facts he had heard, but he wasn't able to *do* anything with them. He lacked the ability to apply in actual practice what he had learned.

The principles discussed in *Waging Business Warfare* work. They have been proved consistently valid throughout history. They have been shaped and reshaped, hammered and altered into their present form, by extremely conscientious men who couldn't afford to be wrong. Every mistake made in battle is paid for in human suffering. There is no margin of error.

But knowing the principles—being able to recite them, for example—counts for precious little. In themselves, the principles are abstractions. They certainly provide a frame of reference for under-standing competitions between corporations and for predicting which competitor will win and where the losing companies went wrong. The lessons and principles also supply a number of key terms for analyzing what happens on the competitive battlefield. Knowing them well, you shouldn't be able to read a business article or be in a meeting without thinking of the lessons and vocabulary of war.

But the information on these pages only springs to life when you apply it. The anthropologist Turney-High may be right. Warfare may be a science, but the application of this science by human beings is an art. Fine generals and excellent managers are wonderful paradoxes. Able to conceptualize effective plans of operation, they can carry them out, too; intensely practical, they are almost without exception contemplative, even studious; aggressively forward-looking, they gain immeasurably by looking back in time for historical precedents; often highly inventive and original, they are constantly borrowing insights and approaches wherever they find them; ceaselessly creative, they rarely let their imaginations get the better of them; concerned with the big picture, their interests often extend to the most minute detail. Above all else, they can translate mere words on pages into action—the one act of management that matters most. In the end, the superior competitor will come out on top, but not without man-agers who both know what to do and are able to do it.

CHAPTER TWO

Leading

The responsibility for a martial host of a million lies in one man. He is the trigger of its spirit.

Ho Yen-Hsi

Hannibal, the Carthaginian (247–183 B.C.), was one of the true masters of warfare. Napoleon considered him superb in every aspect of warring, and Wellington thought him to be the single greatest soldier in all of history. To this day Hannibal's victory against the Romans, commanded by Varro at Cannae in 216 B.C., is considered the most perfect tactical battle ever fought. Of the 76,000 Romans who engaged Hannibal's army of fifty thousand at Cannae, all but six thousand fell on the field of battle.

How could a much smaller army beat a larger one, and so decisively? The historian Polybius provides an important answer: it wasn't Hannibal's soldiers or order of battle that made the Carthaginian army superior to the Roman, according to Polybius, but Hannibal's personal skills. Then Polybius makes a matter-of-fact comment that carries immense implications for corporations vying for competitive excellence: "As soon as the Romans found a general who equaled Hannibal in ability, they immediately defeated him."

In short, a contest within the contest between the Carthaginian and Roman armies was the competition between leaders. The better

23

leader won; the less capable leader lost. More than 125,000 men took part in the battle of Cannae, each pitting his abilities against his counterpart on the other side. Yet it was the qualities of just two human beings—Varro and Hannibal—that dominated the day. It has been said that the mind of the military leader is passed on to ten thousand subordinates. Similarly, Chester Barnard, who was both an organization theorist and the head of the New Jersey Bell Telephone Company, observed that if the president of a telephone company ordered two telephone poles moved from one side of a street to the other side, 10,000 decisions of 100 people at 15 different locations would be involved.

Napoleon also pointed out the significance of the person in charge when he wrote, "It is not the Prussian army which for seven years defended Prussia against the three most powerful nations in Europe, but Frederick the Great." We saw the same thing in chapter 1. The Israelites had the ability to rid themselves of the Midianites all along, but it wasn't until the right leader came along and took charge that they actually did it.

The situation is precisely the same in business warfare. Many corporations have had the ability to gain the competitive edge but weren't able to do so until the right leader with the right strategy, right plan, and right insights into how to manage people took charge and turned things around. We should guard against becoming so accustomed to discussing competitions between corporations that we forget that companies don't run themselves; people run them. We shouldn't forget for a moment that behind the corporate names we will see in this book—GE, Procter & Gamble, IBM, McDonald's, Toyota, GM, etc.—and behind all the strategic and tactical moves we will witness are human beings, the corporate leaders who aren't only managing but pitting their quality as leaders against the abilities of competitors' leaders. French colonel Ardant du Picq (1821–70) made highly detailed and scrupulous studies of the factors leading to success in battle. His most fundamental conclusion was that "it is the mind that wins battles, that will always win them, that always has won them throughout the world's history."

Chief Executive Officers (CEOs) seem to know how integrally leadership ability bears on the well-being of their corporations. When three hundred CEOs around the world were asked recently what they would look for in their successors, "personal leadership style" turned out to be the most sought-after attribute. "Aggressive competitive outlook" was second, and an "entrepreneurial flair" came in third. *Who* a leader is may be more important than what the company is. For example, when the American Can Company's former chairman William Woodside hired Jerry Tsai, Woodside said, "I was looking for a man more than a company. I needed someone who could take five hundred million dollars and quickly generate a presence in a new field." The financial-services company Tsai headed now accounts for 60 percent of American Can's operating earnings. In April 1986 Tsai was made CEO of American Can. Said one observer, "Most corporate executives I've met are sort of pansies and mealy mouths. But Jerry Tsai is a real straight shooter."

"Every successful business is a reflection of one man, one leader," said Armand Hammer of Occidental Petroleum. In his *Art of War*, Sun Tzu (400–320 B.C.) put the issue quite simply. He stated that any commander will be able to forecast which side will win by answering a few basic questions. One of them is "Which of the two commanders has the most ability?"

A GENIUS FOR BUSINESS WARFARE

"Every special calling in life, if it is to be followed with success, requires peculiar qualifications of understanding and soul," writes Clausewitz. Where those "peculiar qualifications" are of a high order and reveal themselves in "extraordinary achievements," the person possessing them can rightly be said to have a genius for that calling. What are the personal qualities of the person with a genius for war?

British field marshal Wavell (1883–1950) claimed that the first essential quality of the leader was "robustness, the ability to stand the shocks of war." Napoleon said the same thing: "The first quality of a commander is a cool head, which will judge things in a true light; he should not let himself be dazed by good or bad news." The Duke of Wellington certainly had this quality of grace under pressure. When told that his critical attack at Toulouse had completely failed, he declared calmly, "Well, I suppose I must try something else." He did and won. Other inner, subjective qualities Wavell considered vital to the leader are knowing what he wants and having the determination to get it. He should have the competitive spirit, the fire inside, and the will to win, even if winning means overcoming great obstacles.

Napoleon said, "If the art of war consisted merely in not taking risks, glory would be at the mercy of very mediocre talent." The person with a genius for war is willing to take the necessary chances. He has, as one soldier noted, "A touch of the gambler in him." What Confederate general Robert E. Lee said about military operations could also be said of running a corporation or a department—there are always hazards in moving, but the possible losses from inaction have to be compared with the risk of action. Usually the risks of inaction are greater.

The person with a genius for war should also possess common sense, the sense of what's practical, possible, doable—based on knowledge of the situation, the competition, and his or her own work force.

To answer his own question—"What is the person with a genius for war like?"—Clausewitz uses an approach that in corporate management is called "situational leadership." He looks first at the environment in which war always moves, then matches environmental factors with personality characteristics. He states that warfare always involves four factors: danger, physical effort, uncertainty, and chance. The leader who functions best of all in this kind of environment— the genius—possesses inner qualities that equip him to deal with those factors. The genius for warfare is aggressive and physically energetic, bold and firm, and has a steady character and calm mind that's unfazed by the circumstances of the moment.

Other military masters have identified the subjective qualities of the person with a genius for warfare. Still other masters of the art haven't put down their ideas on paper but have revealed them in action.

A WILLINGNESS TO COMPETE EVEN IN THE FOG OF WAR

The Spartans are not wont to ask how many the enemy are, but where they are.

Agis II of Sparta, 450 B.C.

In their popular book *In Search of Excellence*, authors Thomas J. Peters and Robert H. Waterman, Jr., describe eight attributes that characterize most nearly the excellent, innovative companies they studied. They called the first attribute "a bias for action," for getting things done: "Even though these companies may be analytical in their approach to decision making, they are not paralyzed by that fact (as so many others seem to be). In many of these companies the standard operating procedure is 'Do it, fix it, try it.' "

The same kind of action orientation is characteristic of the military masters. Rather than "a bias for action," this attribute might best be called "a willingness to fight"; or in business warfare "a willingness to compete," to mix it up with rival companies.

In *every* instance of great tactical success in battle, the person in command demonstrated it—a willingness to fight. No substitute exists for the initiative and resolve that exploit the other side's error or its blindness or susceptibility to deception; or that sees and quickly seizes an opportunity that appears totally unexpectedly or that brings the opponent to battle at the time and place you desire. "War," said Stonewall Jackson (1824–63), "means fighting . . . The business of the soldier is to fight . . . to find the enemy and strike him; to invade his country, and do him all possible damage in the shortest possible time." In all warfare, including the warfare for markets and competitive superiority, you cannot be totally successful if you take a passive stance. Aggressiveness pays dividends.

General Electric is an aggressive combatant in business warfare. Its aggressiveness starts at the top. A former executive with GE said of its chairman, John F. Welch, Jr., at about the time Welch negotiated the purchase of RCA Corporation for $6.3 billion, "If you put Jack in charge of a gas station at a corner with four gas stations, he wouldn't sleep until the other guys had plywood over their windows." "Competitiveness means taking action," says Welch.

I met Ernest Gallo almost thirty years ago when he took the time to describe to this then young visitor to a trade show the qualities of good wine. Gallo is also an aggressive company, and it's looking for scrappy people. It recruits MBAs who are creative, compatible, and dominated by "a sense of urgency." Gallo moves fast and expects the same of its people.

There is another element to the corporate leader's willingness to compete. It's being willing to do so even while knowing that all the information isn't in. It's easy to make your decisions and move when you have all the information you could possibly want. But this rarely, if ever, happens in war or business. Three-quarters of the factors on which decisions are made, says Clausewitz, are obscured by the "fog of war." According to Roy E. Moor, former chief economist at the First National Bank of Chicago, the fog of corporate war is only slightly less thick. According to Moor, approximately 65 percent of the factors affecting a company's marketing success and stock prices are completely outside its own control. Economist John Maynard Keynes pointed out that many business decisions and actions are not based on coolheaded calculation at all but on faith and optimism. He added that if entrepreneurs relied on mathematical equations alone, business enterprise would completely disappear. "Whole brain" research shows that the left half of our brain is rational; the right side, intuitive and creative. Great military masters use both sides in tandem. So does Norman W. Brown, CEO of advertising firm Foote, Cone & Belding Communications. To produce more creative advertising, he teams "left brain" account executives with "right brains," the creative, artsy staff.

Leaders' actions must be taken in spite of the fog of war, and something must always be left to chance. "You've got to do something," said Civil War general William Tecumseh Sherman. "You can't go around asking corporals and sergeants. You must make it out of your own mind." Mediocre staff must be led; superior people need only be directed. But there has to be a clear direction that everyone is pointed in. An executive of AT&T put it well: "We need someone who can give us a rallying cry—throw a stake in the ground and tell us this is what we're gonna do." C. Robert Powell is attempting to turn Reichhold Chemicals Incorporated into a high-tech specialty chemical company. He says he feels a little like a general who finds himself surrounded. The troops ask, "What should I do?" There's just one answer, says Powell. "Attack."

It's only in theory that managers and generals don't move until things are ready and everything has been completely thought out. In reality, the good ones often get things going well *before* the fog of business war has cleared. They seem to know that in most important matters taking a moderately satisfactory course of action right now is usually preferable to taking a more ideal action in the future. Patton, Caesar, and Napoleon were noted for advancing before their supply lines were established. Napoleon said, "I engage and after that I see what to do." If we turned this tendency into a prescription for corporate action, we would have this advice: advance when you're ready enough; never hope to be completely ready—100 percent— for if you are, the opportunity has probably already passed. Numerous corporations are taking this insight very seriously and winning impressive competitive victories. For example, in chapter 5 of *Waging Business Warfare* we will see that Kimberly-Clark is just one company that is attacking giant Procter & Gamble's strategy of entering a market slowly and conducting extensive marketing tests. P&G takes time to see that its product is exactly right while its competitors rush in, at times reaching the market with their product in half the time it takes P&G, and reaping substantial profits.

Performance improves the moment activity and movement begin, and not a moment before. Look around at your work force. If it's withering away in confusion and inaction, it's because of weak and unclear commitments. To regenerate it, commit it to action.

A DECISIVE STYLE OF LEADERSHIP

It is better to live one day as a lion than
a hundred years as a sheep.

Italian proverb

Weak or passive leadership eventually results in the loss of resolution and ultimately brings failure. Strong, decisive leaders bring the opposite. This is one of the fundamental lessons of leadership in war.

Philip of Macedon said, "An army of deer led by a lion is more to be feared than an army of lions led by a deer." Believing this, he raised a lion, his son, Alexander the Great. Hamilcar, the Carthaginian, boasted that he had specifically trained his three sons, Hannibal, Hasdrubal, and Mago, as three lion's whelps, to prey on the Romans.

It's not necessary to be a lion; a bulldog will do. Ulysses S. Grant was an abject failure most of his life. Even his mother called him "Useless Grant." He tried his hand at a number of careers, eventually finding his true calling in war. Grant's leadership style was anything but flashy. "The great thing about him," said Abraham Lincoln, "is cool persistency of purpose. He is not easily excited, and he has got the grip of a bulldog. When he once gets his teeth in, nothing can shake him off."

We don't have to share Mao Tse-tung's political views to admit that he was a leader-general of note. "History," he states, "has never known an infallible general." He cites five major mistakes leaders often make. Each of them is also relevant to business warfare, and each results from indecisive leadership. They are:

1. *Piecemeal reinforcement*, resulting from taking the competition lightly.

2. *Absence of a main direction of attack*. The failure to concentrate—to focus.

3. *Lack of strategic coordination of forces*.

4. *Failure to grasp strategic opportunities*.

5. *The failure to exploit an advantage to the fullest*.

BOLDNESS

"Safety first" is the road to ruin in war.
Winston Churchill

General Sherman defined courage as a sensitivity to the danger involved in an enterprise and "a mental willingness to incur it." "Hesitation and half measures lose all in war," said Napoleon. Machiavelli advised his Prince that "it is better to be impetuous than cautious"; and Gen. Dwight D. Eisenhower returned to civilian life speaking of the value of "living dangerously." It's clear that there are two basic styles of waging war: one is the slow, methodical style; the other is the audacious. The scale of history tips in favor of the second. The story goes that two corporate managers look at the same proposal. One asks, "How much is this going to cost?" The other asks, "How much can we make with this?" Most of the great masters of the military art would ask the second question.

Napoleon's dazzling campaign and battle at Austerlitz (1805) rank among his most remarkable feats. It's often called his most famous victory. Finding himself in a rotten strategic condition—deep in enemy territory, his line of communication threatened, and facing a powerful Austro-Russian army out for revenge for earlier defeats—

his response to the situation was typically brilliant, typically bold. Instead of retreating, the safest course, he turned and successfully attacked the enemy's strongest point, the army at Olmutz, making use of elaborate deceptions that blinded the enemy to his true plans.

Boldness isn't something you save for when you're winning. It's precisely when things are going particularly poorly for you, says Napoleon, that boldness on your part may win the day. If an ordinary general occupying a poor position is surprised by a superior force, he will "seek safety in retreat," writes Napoleon in his *Military Maxims*; but a *great* general "displays the utmost determination and advances to meet the enemy. By this movement he disconcerts his adversary; and if the march of the latter evinces irresolution, an able general, profiting by the moment of indecision, may yet hope for victory. . . ."

Fortunately for Abbott Laboratories, its CEO has more of the great general than the ordinary in him. He became bolder under adversity. Under Robert A. Schoellhorn, the $3.4 billion health-care corporation has clearly outdistanced its competitors. During the period from 1975 to 1985 its 112 percent average total return to shareholders ranked first among the eighteen pharmaceutical companies. Its earnings are double digit, while competitors are hard pressed simply to survive. The key to Abbott's success is its strategy of concentrating on its diagnostic division, which Schoellhorn managed when he joined Abbott in 1973. Beginning, then, with almost no market share in diagnostics, Abbott currently is the world leader. Its sales in the five-billion-dollar industry are 150 percent of those of its nearest competitor.

In 1973 Schoellhorn optimistically predicted annual growth rates of 15–20 percent in diagnostics. But after two major declines, one in 1976, the other in 1978, the sentiment to reduce funding and cut losses ran high. "Bob takes a position quickly and sticks with it," say competitors. "He has strong inner convictions." Rather than cutting back R & D funding when profits fell—as less extraordinary executives might have—Schoellhorn poured *more* money in. Napoleon liked his officers to have a touch of the gambler in them. Schoellhorn does.

Schoellhorn and Abbott showed the winning gambler's touch again in the frenzied race to develop a blood-screening test for AIDS. Its competitors were a full three months ahead of it when Abbott started its research; yet it was Abbott that less than eight months later succeeded in reaching the marketplace first with the test. The thirty-person team assigned to the project was given these instructions: "Don't let anything stand in your way."

At Chancellorsville (1863), the victory of southern forces, commanded by Robert E. Lee and supported by Stonewall Jackson (one of the greatest fighting teams in history), over those of the North under Maj. Gen. Joseph "Fighting Joe" Hooker was not a result of numerical superiority. The South was outnumbered two hundred thousand to sixty thousand. The deciding factor was solely Lee's willingness to take a chance on bold, aggressive action. He ordered Jackson to march right across the enemy's front. On the other side, Hooker was mired in overconfidence and indecision. Not only wasn't Hooker decisive; he didn't have the most potent weapon you can possess as a leader—the hearts and minds of your staff. Hooker was anything but respected by his men. They thought he was arrogant and that he overindulged in liquor and women. Demonstrating their low opinion of him, they termed the prostitutes that followed the army "hookers," a tag that's stuck.

The argument could easily be made that boldness in and of itself is what makes winners in business warfare. Legion are the examples of companies that grew from humble beginnings into huge diversified corporations while competitors stayed small or completely disappeared. Invariably it will be found that the winner was willing to take risks that the loser feared to take. In its earlier days, Texas Instruments gained the competitive edge precisely because of its boldness and daring. Though it was a small company, it outcompeted giants like RCA, Bell Labs, and GE in the semiconductor industry

because it was willing to take its products quickly out of the labora-
tory and into the field—to go out and try to *do* something. One small
company that's showing the boldness of the type that can lead to
significant size is Clothestime Incorporated, a rapidly growing young
women's clothes retailer. The company found a profitable niche when
department stores began shifting away from adolescent clothing to
career women's styles. Clothestime is moving so fast that it once
opened twenty-one outlets in a single day, adding 10 percent to its
size.

Sherman said, "You cannot attain great success without taking
great risks." Trammell Crow, commercial real estate developer, is
just one entrepreneur whose career demonstrates the truth of Sherman's words. Crow started small, took enormous risks, and is currently
the country's biggest landlord, with office buildings, warehouses,
apartments, hotels, motels, and shopping centers across America. Not
long ago, his firm, Trammell Crow Company, was cited as one of the
ten best companies to work for in America.

A study by R. Joseph Monsen described in *Business Horizons*
looked at differences in policy and performance between companies
controlled by their owners and those in the same industry but in the
control of professional managers. Companies that were owner controlled had much higher rates of return on investment. One reason
for the greater efficiency of the owner-controlled companies was
thought possibly to be found in "motivational issues"—the owners
had more of a financial incentive for taking risks than managers did.

Companies whose policies reward managers who meet and exceed
goals condition managers to take risks. But companies that punish
managers for failing to reach their goals communicate the value of
playing it safe. Whether warrior or manager, the person who values
self-preservation above everything else will hesitate to take chances.
In many companies one failure can end or retard a career. To preserve themselves, managers in those companies are understandably
inclined to make their decisions at the low-risk-but-low-payoff end of

the decision continuum. William Smithburg of Quaker Oats Company is just one executive who is teaching his managers the importance of living at least a little more dangerously. Smithburg learned the art of marketing while an advertising account executive. Hired by Quaker Oats to rejuvenate its faltering ready-to-eat cereal business, he became known for successfully implementing bold marketing strategies. Now chairman of the company, he established an attitude of increased boldness throughout the company. Says Frank Morgan, Quaker president, "One thing he brought to the company is the notion of reasonable risk-taking. He wants people to take responsibility for their businesses, to set aggressive plans and take chances."

Is age a factor? Is boldness a youthful attribute that disappears as a manager, soldier, or company grows older? Texas Instruments, so bold in its early dates, lost some of its daring with age. The change corresponds with its market decline. Long ago, King C. Gillette, founder of the Gillette Company, took an enormous chance when he went against his advisers and emphasized the selling of blades over razors. Today many critics claim the company doesn't do better because it has become averse to risking. As he aged, Frederick the Great lost some of his zest for fighting, and American automobile manufacturers had become so cautious that when the Japanese came with their new small-car designs, the American companies had trouble responding.

Machiavelli advised the Prince to select young leaders because they were less cautious, fiercer, and more audacious. Napoleon's marshals were astonishingly young; only two were over forty. He believed that young colonels would be more daring in battle, while older generals with careers to protect would invariably choose the safer course. One commentator puts it very bluntly: if Napoleon had more young colonels at Waterloo, he would probably have won. Alexander the Great was a very capable military commander by the time he was eighteen. Two years later he was a king. At twenty-six, an age that today would make him a junior executive in a neighborhood bank, he had beaten into submission the most powerful empire Western civilization had ever seen. At thirty (he would be a bank vice-president now) he ruled almost all of the known world. Without changing a thing—not weapons, strategy, tactics, or leadership style—Alexander could probably have beaten every army to be assembled anywhere on earth during the next eighteen hundred years until guns dominated wars.

Maturity can reduce boldness but needn't. It depends. Surely you've known old-timers—companies and managers—who have never lost their vitality. Napoleon called Turenne (1611–75) "the greatest of the French generals"; the only one "who became bolder with age." Turenne began as a mediocre commander but made himself into a great one through studying hard what brought victory, or defeat, and carefully creating his own style of combat—painstaking preparation and bold, sweeping maneuvers. His enemies were cautioned that when facing Turenne it was dangerous to make a mistake. The career of this self-made man is amazing. It would be as if a corporate manager who was once very average fashioned himself into the best and boldest executive in any company anywhere.

PRUDENCE

A good general not only sees the way to victory; he also knows when victory is impossible.

Polybius, 125 B.C.

Eventually Napoleon succumbed to the disease that destroys leaders —egotism. He came to consider his enemies average or below and himself invincible, the only general in all the world who was fit to lead a great army. He made the terrible mistake of considering Wellington, the man who would beat him at Waterloo, just another overly cautious commander. He failed to realize that Wellington was cautious only when the situation dictated, but that when the advantage was his, Wellington struck like a devastating bolt of lightning. Napoleon wrongly believed that no impediment was insurmountable for him. Consequently, he unwisely undertook actions without sufficient means. Had he not fallen victim to his egomania, he would possibly never have taken his army to Russia, where death awaited it.

Antoine Henri de Jomini, military adviser to Napoleon, remarked of him: "His victories teach what may be accomplished by activity, boldness, and skill; his disasters, what might have been avoided by prudence." You can easily find the same kind of exaggerated self-confidence in business warfare, too.

Dr. Eugene Emerson Jennings, professor of administrative science at Michigan State University, studied the qualities that made for success or failure of corporate executives. He discovered that two noticeable traits of "failure prone" executives are "illusion of immunity" to bad luck and an "illusion of mastery" over life's events. When the unexpected hits, the failure-prone executive, living in a land of illusion, is likely to be knocked off balance by the shock of reality. The more consistently successful executive hasn't deluded himself. When things go against him, he's prepared and able to adjust. Virtually every great leader, unlike the failure-prone executive, takes the unexpected into account and prepares himself and his people for it. Hannibal's center was broken at Cannae. A devastating thing to a lesser man, it didn't faze Hannibal or his soldiers. Before the battle Hannibal had explained his plan to his soldiers, warning that at some point the center would be broken but that they shouldn't worry about that because it was part of his plan.

Thomas L. Berg, author of *Mismarketing: Case Histories of Marketing Misfires*, calls overoptimism in the face of the unexpected a recognized enemy of success in marketing: "Overoptimism stems from a failure to fear failure. It expresses itself in empty and unrealistic bravado, a view of the marketing world through rose-colored glasses."

Boldness stops at the outer limits of probability—or should. Sun

Tzu wrote, "He will win who knows when to fight and when not to fight," and in his *Histories* Polybius wrote, "A good general not only sees the way to victory; he also knows when victory is impossible." The best risk is a calculated one. No matter how brave the master of business warfare is, he or she understands that some options must not be followed, some companies must not be attacked, some markets shouldn't be contested. "Boldness, directed by an overruling intelligence," wrote Clausewitz, "is the stamp of the hero: this boldness does not consist in venturing directly against the nature of things, in a downright contempt of the laws of probability. . . ."

Extensive research has been done on people with a high level of achievement motivation. These people are dominated by the need to accomplish, to get things done. On the whole, they tend to make more money than others, to find jobs more quickly when out of work, and to be promoted more rapidly. Nations that have more of them in their population have a higher gross national product than countries with fewer of them. Peter Engel was simultaneously president and CEO of Helena Rubinstein and corporate vice-president of Colgate-Palmolive when he said that the main concern of any business enterprise should be to hire and nurture people dominated by achievement motivation. While we might expect that these extraordinary achievers take unusually high risks, that isn't the case at all. Typically they choose to pursue reasonably difficult goals; goals that they believe they have approximately a fifty-fifty, or at least a one-in-three, chance of reaching. They take chances, but not chances that in Clausewitz's terms defy the laws of probability. Here are two important truths of warfare pertaining to prudence.

- Move forward if it's to your advantage; if it isn't, stay where you are.

- Never hesitate to commit yourself to battle to extricate yourself from some serious difficulty or to put an end to a competition that otherwise might ruin you. The basis of strategy is common sense, and common sense tells you to fight for your life.

CHARACTER

*He must have "character," which means that he knows
what he wants and has the courage and determination
to get it.*

Sir A. P. Wavell

The Greek philosopher Heraclitus taught that "a man's character is
his destiny." We can go further: a corporate leader's character is
probably his or her company's or unit's destiny. The last person you'd
want to follow into business battle is a weasel or a coward. Warfare
offers the following lessons about the character of effective leaders:

Great leaders are highly individual. Leaders of competitive enterprises do have certain attributes in common—the willingness to
compete, for example, and a decisive style of leadership. But no two
are mirror images of each other. The Duke of Marlborough (1650–
1722), whose victory at Blenheim (1704) shattered the myth that
the French were invincible, was loved by his men for his sweet and
serene personality. But Sherman was a master, too, and he was
thought by almost everyone close to him to be mad as a hatter and
addle brained. Once in battle, however, his idiosyncrasies seemed to
disappear. His mind became sharp and clear. Alfred P. Sloan of
General Motors has been called the epitome of "the organization
man." Yet, ironically, on the personal level he was a loner who had
few close friends, perhaps not one.

Edwin E. Ghiselli's studies of managerial talent reveal that those
who show the greatest individuality in managerial behavior in corporations in general are judged to be the best managers. Lee Iacocca
is a folk hero. He enjoys being in the public eye, but the man who
doubled Bendix Corporation's sales to three billion dollars in just
five years—W. Michael Blumenthal—is just the opposite. He shuns
publicity, working quietly to bring his present company, Burroughs,
on a more competitive footing with IBM. It was Blumenthal who
engineered Burroughs' $4.8 billion acquisition of Sperry Corporation
to form Unisys Corporation in November 1986, and position Unisys

as second only to IBM among U.S. computer companies. A short nine years after borrowing twenty-five thousand dollars to start his first business, William F. Farley acquired conglomerate Northwest Industries for $1.4 billion, one of the biggest leveraged buyouts in business history. Farley, a fighter and anything but eccentric, is nevertheless highly individual. A devotee of fitness, he begins his management meetings with an hour of aerobics. He has managers sing fight songs.

Even though operating in precisely the same industry, two corporate leaders may have polar-opposite personalities. For example, Charles Schwab, one of the most recognizable faces and names in America, is extroverted and flamboyant. The discount brokerage company he heads, Schwab & Co., is located in a fancy new skyscraper, the Charles Schwab Building in San Francisco. Everything is first-class. On the other hand, Schwab's chief competitor, Quick & Riley Group, Inc., is headed by Leslie C. Quick, Jr. No one has ever heard of him, yet *Business Week* recently called Quick "probably the wealthiest CEO on Wall Street." Though running a company worth $120 million, Quick is cost conscious to the extreme. His "world headquarters" consists of fifteen people, all of whom have offices outfitted with used furniture, the only kind Quick will buy.

Even the same corporation can function under leaders with polar-opposite personalities. Control Data's CEO, Robert M. Price, has a background in mathematics and trusts the numbers. His predecessor, William C. Norris, who operated the company for twenty-four years, trusted intuition. When AT&T replaces chairmen, it appoints one with a personality totally opposite that of his predecessor.

Here are some additional truths of warfare pertaining to character:

- If the principal leader believes the plan he is asked to follow is a bad one and will lead to defeat, he should oppose it.

- The best leaders are objective about themselves; they have the strength of character to recognize their limitations.

- Bullies and sadists corrode the character of their subordinates.

- Scapegoating is always a sign of weak, spineless leadership. When scapegoats are brought to sacrifice for every conceivable mistake, said Erwin Rommel, the Desert Fox, "It usually shows something wrong at the very highest command. It completely inhibits the willingness of junior commanders to make decisions."

- If you're afraid of unfavorable short-term repercussions, you'll have to pay the price in the long run. Better to have the fortitude to stand up to criticism now.

- The quality of your character is revealed in every important decision you make.

- Your character is measured by the strength of your convictions.

- The more intensely you're involved in the action of business warfare, the more susceptible you will be to self-doubt and the temptation to change your plans.

- Courage in the face of fear is part and parcel of effective leadership character. In his studies of emotional problems of managers and engineers, Michael Maccoby discovered that half the people interviewed cited anxiety as a problem for them and almost one-third felt "unwarranted fears" were a problem. Fear of not performing well, losing a sale, missing a deadline, being wrong, looking bad, getting fired, losing control, are just a few business fears. Clausewitz identifies two kinds of courage: (1) physical courage—when you, your status, or career are threatened; and (2) spiritual courage (*courage d'esprit*)—the courage to face up to your responsibilities. If you're courageous in both combined, says Clausewitz, you possess "perfect" courage.

- The strength of character you display creates strength in your subordinates, particularly if you show it when you and they are under great stress. When a young IBM executive took a risk that turned out badly and cost the company ten million dollars, IBM president Tom Watson, Sr., called the man into his office. The young man said, "I guess you want my resignation." Watson replied, "You can't be serious. We've just spent ten million dollars educating you!"

- Managers and executives who bail out under pressure don't deserve to ride in the same elevator as those with character who see things through to the end. The military epitome of the latter was Ulysses S. Grant. Grant-type managers are the kind who—no matter how badly things are going—hang in there and fight.

- The man or woman of high character is ready to accept the blame when things go wrong. Hardly an executive lives who doesn't hope for more honesty from his subordinates—at least a few honest men and women.

- Personal integrity, dependability, and equitable treatment of subordinates are indispensable elements of sound leadership. Integrity to stand behind what you believe is right; dependability in demonstrating that you will perform the job required of you; equitable treatment in rewarding and punishing performance consistently and impartially. "A prince," wrote Machiavelli, "should show himself a lover of merit."

GREATNESS AT THE CRITICAL MOMENT

He was above all great in the most critical moments.

Napoleon, of Frederick the Great

What distinguishes leaders from the crowd is their talent for coming through "trailing clouds of glory" at the critical moment when a crucial issue hangs in the balance. "At the critical moment," writes Sun Tzu, "the leader of an army acts like one who has climbed up a height and then kicks away the ladder behind him."

The surest path to leadership stature is by overcoming great adversity. As Lee Iacocca's Chrysler experience demonstrates, if reputation were your only concern, you could hardly hope for anything better than a huge problem to solve or a crisis to overcome. Whenever Fortune wants to render a new prince great, says Machiavelli, she raises up enemies against him, compelling him to do battle with them, so that by overcoming them, he ascends to greatness.

James MacGregor Burns makes the same point in his Pulitzer Prize-winning book *Leadership*. He describes a number of types of leaders, including the "heroic" leaders. These leaders are not believed in because of their abilities or track record or their stand on issues but because of "their personage alone." They are believed in because of who they are. Heroic leaders, Burns adds, in contrast to leaders who are merely enjoying popular favor, usually emerge when there is a profound crisis. Machiavelli suggests that a wise prince might consider fomenting enmity if it doesn't already exist—to create an enemy, suppress it, and thereby become great.

General Custer didn't come through at the critical moment. Little Big Horn was his critical moment, and when it came, he failed. That really shouldn't surprise anyone. Here was a man who was beaten at will by the Confederates and who never joined President Lincoln's victory parade because his horse ran away from him.

When you take over a company or unit, you've reached a critical moment, particularly if you're succeeding a manager who was well liked by his or her staff. In August 1792 French general Charles Dumouriez replaced such a man. As soon as he had arrived, Dumouriez assembled his new army. Instead of the traditional cheering, he was confronted by a surly bunch. One soldier shouted from the ranks, "It's that bastard who declared war."

Dumouriez yelled back, "Do you think that liberty can be won without fighting?"

When another cried out, "Down with the general!" Dumouriez drew his sword and challenged him to a fight. When the man slid back, the general realized that his unorthodox introduction had won his army over. Their respect and hearts were his.

Corporate leaders have been known to employ dramatic Dumouriez-like gestures to signify something important. For example, one early morning in June 1985, Texas Instruments' president and CEO Jerry Junkins took off in a company plane to visit company plants in the middle of a storm. He was trying to communicate that TI can meet tough times and come out alive.

August Busch III (called "Little Augie" in St. Louis) *became* great at the critical moment when he succeeded his father as the head of Anheuser-Busch Companies in the mid-seventies. "*Became*," because although he turned out to be a great leader, transforming a loosely managed corporation into a finely meshed, increasingly dominant force in the brewing business, there was little in Busch's background that foreshadowed greatness. In his younger days he was, according to a *Business Week* article, "an aimless playboy." Such sudden transformations from mediocrity to great leadership performance are not nearly as unusual as one might think. Years before he became president of Neiman-Marcus, David Dworkin was unloading trucks in Florida when something like a religious experience persuaded him that he was destined for some kind of greatness. He had a picture of himself being in charge of a company. He quit his job and hopped a plane for New York. The point being, be very careful judging a person totally on the basis of his or her past.

A man who was about to head a faltering company with terrible morale happened to telephone me when I was working on this section of *Waging Business Warfare*. "Whatever you do," I said, "be great at the critical moment. Do something significant right away." Learning that the workers had never been asked to voice their views on anything, ever, he got them all together in an auditorium and asked them as a group to list the qualities they would like in a CEO. They itemized such things as dedication, willingness to listen, flexibility, enthusiasm, trust, etc. When they had finished, he looked at the list and said, "You got it." They cheered.

Often new CEOs signal the changing of command by declaring daring goals, only to find that soon everyone simply goes back to business as usual. The only goals that will be followed, however forcefully presented, are those that the workers believe are realistic.

In describing "the slow disturbance of equilibrium which commences soon," Clausewitz is pointing out that one side gains the advantage and does so *soon* after the battle begins. Once gained, an advantage is usually held: "Certainly there are instances of battles which after having taken a decided turn to one side have still ended in favor of the other; but they are rare, not usual," says Clausewitz. Thus, the moment immediately after launching a major effort is a critical one.

Another key moment happens immediately after the first engagement with the competition. Things may end in a draw but usually don't. Ordinarily one side wins, one loses. Your company wins the big Department of Defense contract, or the competition gets the business; or they get the product out into the market while your version is still on the drawing boards. The winning side will be confident and in high spirits immediately after the victory. Then—at that point of victory—the winning side has reached a critical moment. The art of war's subprinciple of pursuit (see chapter 5) states that no win is complete unless the issue is pressed forward and the opponent is driven to greater loss. Absurd as it seems, this rarely happens. Relentless pursuit of an advantage once gained is one of IBM's secrets of success in computer wars, but IBM is unusual. Pursuit is probably the most often violated truth of effective warring. At this moment the task of the corporate leader is to spur the work force on to gain even more of an advantage.

The confidence of the side losing the first engagement will be lower than that of the winning side is high. The effects on morale are always more profound on the beaten than on the victor. Staff will fall victim to self-doubts: "They're better than we are." The loss of the competition often leads to the loss of confidence in the leader: "The feeling of being conquered . . . now spreads through all the ranks, even down to the common soldiers . . . aggravated by a rising distrust of the chief commander." Col. Ardant du Picq was one of the really brilliant and original students of war. He observed that the more confident the losing side was of its plans and projects, the more disillusioned its soldiers will be to see that the best they could do was not good enough to beat their opponent.

The manager who's great at the critical moment immediately after a loss will try to dispel the worker's low spirits as quickly as possible. In 1974, when Edward J. Noha became president, chairman, and

CEO of CNA Insurance, the company had reported almost a two-hundred-million-dollar loss. Some analysts were predicting that CNA's chances of surviving were nil. "It might have been stupidity or ego," said Noha, "but . . . I never for a moment thought this place would not be around. . . . I figured I could turn this baby around." While Noha was confident, insurance insiders weren't. Some, waiting for the other shoe to fall, fully expected one of the biggest insurance-company flops in the history of American business. Noha felt that his first task was to restore confidence in CNA—the confidence of CNA's twelve thousand employees as well as public confidence. Over the next three months he walked through all forty-four stories of the CNA building and introduced himself personally to every employee.

Since those tough days Noha has brought about the resurgence of CNA. He took other steps to put the company back on a sound footing, but perhaps none as significant as that single gesture of great leadership at the critical moment. "People to this day remember that," he said.

To rebuild the fallen confidence of workers when the company has experienced a setback, it's always wise to:

- Look for the root causes of the setback and remedy those that you can.

- Express your confidence in employees' abilities.

- Inform them of the situation and the measures the company is taking to regain the initiative. It's always preferable to be open and aboveboard in admitting difficulties than to try to hide the facts.

- Disseminate information on all the things going right for the company—profits being made in other divisions, for example.

- Personally visit units of the company.

- If the setback is a result of being beaten by a competitor, Frederick the Great's advice is worthwhile: don't despise the enemy, but never speak of him to your own people without expressing your confidence in your side. Carefully point out the advantages of your employees over theirs.

Unless it's absolutely impossible to do, it's always advisable to regain your army's spirits by engaging the enemy that has beaten you—and as quickly as you can. Lee beat Grant in the campaign against Richmond, but Grant, new to the command of the Army of the Potomac, did what no other Union commander had done after being bested by Lee. His predecessors had invariably retreated a great distance. On the evening after being beaten, Grant positioned himself in the road. As each defeated regiment approached, Grant motioned for it to take the right fork in the road—back toward Lee's flank.

A sudden turn of events is another critical moment—as when a company has been following a stable strategic course for years and suddenly finds the environment has changed markedly. Most corporations will eventually adapt, but before they do, their managers, like a quarterback who has been sacked hard, will for a time be stunned, confused, their corporate bells rung, and wondering, *Where did we break down?* At such moments it's always wise to take your mind off your problems for a while and think what a rotten time your competition must be having.

GRACE UNDER PRESSURE

... cool, yet daring in the midst of peril ...
E. S. Creasy, *Fifteen Decisive Battles of the World*

The shocks of business warfare are many and varied—the market changes, or your new product fails, or you suddenly develop a cash-flow crisis. Morton Thiokol's aerospace group experienced monumental shocks when questions were raised about its involvement in the disastrous explosion of the Challenger spacecraft. Johnson & Johnson endured the shocks of war when their Tylenol capsules were poisoned. No one has been able to substantiate even one of the wave of complaints in 1986 of glass in Gerber baby food, but the rumor alone has been sufficient to damage, at least temporarily, Gerber's reputation for safe, quality products. James Lovejoy, Gerber's director of corporate communications, says, "Nobody wants to wait for due process. We're guilty until proven innocent." The collapse of the London tin market cut American Express Company's 1986 first-quarter net income up to forty-seven million dollars. And in addition to these examples of corporate pressure are the less dramatic everyday pressures that take their toll on every manager. Stress, studies estimate, cost American corporations approximately $150 billion each year.

Clausewitz calls "a strong mind," one that in combat can become excited, yet maintain serenity under that excitement so that perception and judgment remain free, "like the needle of the compass in the storm-tossed ship." Most great leaders have had this inner strength of mind that enabled them to withstand the shocks of war with style and class. Even when deepest in trouble, Franklin D. Roosevelt never exhibited more than a mild exasperation. General Motors chairman Roger Smith has been called the most innovative person in that job since Alfred Sloan. Intensely involved in restructuring GM and innovating and diversifying broadly, he still appears to be "light-hearted and having a wonderful time in his job."

The Duke of Wellington won every battle he fought. It was his personal qualities, even more than his careful planning and bold execution, that consistently brought him victory—in particular his calm courage, decisiveness, and coolness at the moments of extreme pressure. At the Battle of Salamanca in Spain, July 1812, a British army of fifty-two thousand under Wellington (1769–1852) faced forty-seven thousand French troops. When Marmont, the French commander, noticed the dust raised by the march of the British Third Division, he assumed that Wellington was preparing to move off in that direction and ordered the leading French divisions to shift to the left. For some reason the two French division leaders became confused and started to race their troops, throwing all caution to the wind.

Wellington was eating lunch when told of the French move across his flank. Immediately he seized a telescope and, after a quick look, exclaimed, his mouth still full, "By God! That will do!" Then he leaped on his horse and galloped over to the Third Division, going directly to Pakenham, his brother-in-law. "Ned, move on with the Third Division, take those heights in your front, and drive everything before you." He then jumped back on his horse and raced back to the center of his line. Throughout his famous "forty minutes," the most critical moments of his career to date, Wellington raced from place to place on the battlefield giving decisive orders in rapid fire to one division after another. In less than one hour—forty minutes—the entire left wing of the French army was routed and its commander dead.

It's an odd fact, but it has rarely happened that the two greatest generals of any era have faced each other in battle. So it's remarkable that three years after Salamanca, Wellington's opponent at Waterloo (1815) was Napoleon. Waterloo, incidentally, is the battle that tough, competitive Pennzoil chairman Hugh Liedtke enjoyed reenacting with toy soldiers when he was a child.

It was 3:30 in the afternoon when Napoleon decided to try to crush the British center and right by hurling his magnificent heavy cavalry against them. But Wellington had formed his infantry in squares, and each French charge proved futile against the solid wall of British bayonets.

Throughout the battle Wellington raced to the critical point and encouraged his men. To a battalion that was being hit particularly hard by French artillery, he said, "Hard pounding this, gentlemen: we will try to see who can pound the longest." "Stand firm, my lads; what will they say of this in England?" he remarked to troops in a position that had seen furious action and was expecting another attack. One English unit had been shelled for a long time and ordered to wait. Tired of being shot at, they wanted to *do* something. Had they moved too soon, the Battle of Waterloo would probably have been lost. Wellington raced over to them and called for patience: "Wait a little longer, my lads, and you shall have your wish." The troops nodded and stood fast.

When any of his generals begged him for reinforcements or for permission to retreat, his answer was always the same. "It is impossible; you must hold your ground to the last man, and all will be well." When his staff asked him what his plan was so that it could be carried out even if he fell, he replied, "My plan is simply to stand my ground here to the last man." At one point he realized that his officers had congregated around him and that the French batteries were aiming their shots at the group. Always cool, he said, "Gentlemen, we are rather too close together—better to divide a little."

Later, when he saw the confusion the French had fallen into, he decided the time was right to finish things. He spurred his horse, Copenhagen, to the edge of a plateau, took off his hat, and waved it in the air. At the signal, forty thousand English troops poured down from the slopes they had occupied. The enemy's line began to yield, then broke. A loud cheer went up from the British soldiers, and soon Waterloo was history.

WILLPOWER

... As soon as difficulties arise ... the commander must have a great force of will.

Clausewitz

Alexander the Great *was* great for many reasons, none more significant than his demonic strength of will. Said French general Ferdinand Foch, "Victory is a thing of the will." Often when everyone else is giving up, the entire enterprise may come to rest on the will of the leader. This may be the make-or-break moment—for the leader and the corporation. A marketing executive for a once-great but now mediocre household-appliance company described the devastating effects of the CEO's loss of will. "We relied on Don so much, and he was always so strong. When he lost his confidence, we all lost ours, and all was lost." Studies reveal that materially successful people are not necessarily any more intelligent or gifted than those who don't do as well. *Perseverance*, a quality of the will, is the one factor that distinguishes the highly successful from the less successful individual. Often the victor wins by simply sustaining the will to win longer than anyone else. No general was more remarkable for maintaining the will to win than the Confederate John B. Hood. At Gettysburg one of his arms was crushed, and at Chickamauga he lost a leg up to the hip joint. He ordered himself strapped to his saddle and continued to lead his troops.

Gail Borden, founder of Borden's milk, was fifty-six and on what he called the "downhill side of life" when he developed his first successful business. Before achieving success, this man of immense willpower experienced one failure after another. He started out as a surveyor in his teens and by the age of twenty was county surveyor for Jefferson County, Indiana. He became a schoolteacher in Mississippi but quit that to invent a prairie schoonerlike wagon equipped with sails that was supposed to make it possible to travel by land or water. On its maiden voyage the contraption capsized, pouring a wagonload of prospective buyers into the Gulf of Mexico. He then invented a Lazy Susan that even his wife said was impractical. A failure and broke, he became a newspaperman in Texas, then landed a job as customs collector for the port of Galveston. Back to inventing, he developed a useful dehydrated meat biscuit, but no market developed for it, and he was broke again, though the product did receive a gold medal at the London Exposition of 1851.

It was on the return voyage from the exposition that he witnessed something that changed his life. He noticed a number of immigrant infants who were crying in the boat's hold. In a few days four of them died from drinking contaminated milk from diseased cows on board. Deeply moved by their deaths, he vowed to discover a milk that could be safely used on ships. He experimented with "condensed milk" for two years and after a number of battles over the patent finally received the rights to manufacture the product. He opened a factory but once again found that there was no market for his product and had to shut it down.

While riding on a train, he happened to start up a conversation with a New York banker who agreed to back him in his latest venture to the tune of one hundred thousand dollars. When the banker died some years later, his investment was worth eighty times his original investment. When Borden died in 1873, his company was clearly the leading milk-condensing firm in the country, with sales throughout the world. On his gravestone is inscribed the epitaph he wrote: "I tried and failed. I tried again and again, and succeeded."

It's so much easier for human beings to put off the thing that should be done than to do it now; not to do than to do; to throw up their hands and let situations work themselves out than to initiate action and make things happen.

Willpower is that conscientious voice inside that reminds you to resist inertia and to do what you are tempted to put off. The great samurai leader Takeda Shingen gave a piece of advice that initially seems terribly harsh but acquires more sense the more you think about it. He said, "If only a man will not do what he himself would like to do, and do those things that he finds unpleasant, his position, no matter what it is, will be much improved."

INNOVATIVENESS

The unresting progress of mankind causes continued change in the weapons, and with that must come a continual change in the manner of fighting.

Rear Adm. A. T. Mahan (1840–1914)

Almost all the battles considered masterpieces have involved the use of a novel strategy, device, or move on the part of the winning side. Characteristically, the military masters of strategy and tactics seized upon and refined at least one innovation that became a key element of their particular style of fighting: Hannibal's insatiable appetite for the surprise attack (see chapter 6), Frederick's attacks in oblique order (chapter 5), Wellington's attacks from a defensive position (chapter 5), and Napoleon's quick and powerful concentrations (chapter 4), for example. Successful corporations, too, may shape their particular style of competing around a certain innovation. The giant K mart owns twice as many stores as any other retailer. Its chief technology is warehousing, buying related goods in huge quantities at low prices. This technology is not only K mart's chief weapon; it *is* K mart. If K mart abandons this strength and diversifies into smaller-volume, unrelated specialty stores, it will be in trouble. Innovation can completely change the rules of the competition, whether in warfare or business. For example, the technological invention of the simple saddle with stirrups had a tremendous impact on warfare. It gave the horse soldier the stability in the saddle to hold a stout lance under his arm and deliver a heavy blow while also retaining maneuverability.

Innovativeness in business has probably never been as important to survival as it is now, simply because of the quick-fire pace of change. Technological innovation and marketing innovation are two areas that dominate the minds of corporate winners. It's naive to believe that failure to innovate alone kills corporations, but it certainly enters prominently into the formula for failure. For example, International Harvester's rapid decline is often attributed to problems innovating. On the other hand, the survival of Caterpillar, IH's direct competitor, is credited to Caterpillar's incessant efforts to innovate its products.

CONCERN FOR MORALE

. . . passion brings about accomplishment.
King Archidamus of Sparta

John Brodie, former NFL quarterback, said, "There are times when an entire team will leap up a few notches. Then you feel that tremendous rush of energy across the field." You have probably felt that rush of energy—in athletics and at the office, too. When it's there, it's a difference that makes *all* the difference. The entire work force is committed to one objective. I call its opposite "the Shade 44

Purple Paint syndrome" from a story a manager of a paint manufacturer told me. It seems that if an executive from headquarters
touring the manufacturing plant acted arrogantly, he would be taken
to the place where Shade 44 Purple was being made. An "accident"
would be arranged, and some of the paint would be spilled on him.
"Shade 44 Purple doesn't come off," I was told.

Great military leaders and corporate managers alike have placed
the highest value on getting their people to "leap up a few notches"
by turning their hearts and minds toward victory. MORALE is a state
of mind producing determination, zeal, and the will to win, and it,
too, is one of the competitions within the competition of business
warfare. Du Picq states the point that leaders in every competitive
field from the beginning of time have believed: the battle in the field
or in the marketplace is an expression of another struggle—that
between human wills. Whoever is stronger in the head and gut will
generally win. It's been said that three things are needed to carry on
war: money, money, and more money. But Napoleon considered
morale three times more important than money or any tangible object
it could buy. "The moral is to the material as three to one" is probably
Napoleon's most-quoted maxim.

No one was ever better than Charles XII of Sweden (1682–1718)
at getting a team to leap up a notch or two. One commentator has
referred to him as "the most extraordinary soldier in the history of
war." His successors didn't study Charles to learn more about his
methods, for they weren't particularly outstanding. Yet warriors
scrutinized his career for many years after his death. What distin-
guished the man *was* the man. It was his presence that created loyalty
and morale. More loved and believed in than perhaps any leader in
any field of competition, he provides valuable insights for anyone
interested in the effect of leadership on morale.

First, he was not only willing to compete, to fight; he fervently
loved to. Some have suggested that he was even more interested in
fighting than in winning.

A second factor was his personality. He embodied the best qualities of the inner spirit. He was afraid of nothing, and no matter what was happening, he always retained his optimism and good spirits. He was tough and demanding but fair and completely honest with his men. He never asked anyone to do what he wasn't himself prepared to do. Personal courage came easily for him. He seems never to have thought of not showing guts, energy, and total commitment, no matter what hazards lay ahead of him.

Third, he created legends about himself. Many leaders pay very close attention to creating the most favorable image—for their employees, competitors, and the general public—a kind of legendary stature. Charles didn't set out to do legendary things. He just did them. Every action he undertook was "somehow invested with glory." In 1713 he took forty men with him to hold off twelve thousand Turks. His small unit was taken prisoner, but not until it had cut down two hundred Turks. Charles killed ten himself before being taken prisoner.

COUP D'OEIL

There is a gift of being able to see at a glance the
possibilities offered by the terrain.

Napoleon

Warriors refer to "a strategic sense" as *coup d'œil*. Napoleon called it his "star"; Alexander the Great, his "hope"; and Caesar, "luck." For Erwin Rommel, it was *Spitzenfingergefühl*, "feeling at the finger tips." All would agree that it's not equally distributed. Some people you know have a flair for constantly finding opportunities and coming up with damn good ideas, and others seem to pass their entire careers without generating even one.

The *coup d'œil* is the ability some people have of conceiving "in a moment all the advantages of the terrain and the use that they can make of it with their army." It enables you to look over a battlefield or corporate competitive scene and take in immediately the advan-

tages and disadvantages, to relate the parts to the whole and to spot
the most advantageous points of attack. Gustavus Adolphus wasn't a
better cavalry commander than his adversary Pappenheim and perhaps wasn't as good. But Pappenheim didn't have Adolphus's *coup
d'œil* and simply couldn't grasp a whole battle at once.

People who have *coup d'œil* in abundance seem able to predict
what the opponent will do. General MacArthur, for example, could
foretell with detailed accuracy precisely what the enemy's future
course of action would be. Great corporate managers also have this
ability. For example, Samuel J. Heyman, chairman of GAF Corporation, is known for having "unusually good instincts," and it has been
said of Donald Trump, the building baron, that "he has an absolutely
incredible ability to see what's going to happen."

"The thing you learn working with Arthur is that the guy is right
all the time." Arthur is Arthur Rock, the man with the best track
record in the venture-capital business. He operates alone at times, at
times with a single partner, and at times in the high-risk high-tech
industry that most venture capitalists avoid. He always operates with
instinct. While others shied away from a strange little computer
business, Rock sank almost sixty thousand dollars in it. The company
created the personal-computer business and brought a three hundred-
fold return to its backers—Apple Computer. Perhaps the most interesting thing about Rock is that he not only invests from his *coup
d'œil* but *in* that of the leaders of his investment. "I invest in people,
not ideas," he says. "If you can find good people, if they're wrong on a
product, they'll make a switch."

The *coup d'œil* has three dimensions. The first consists of grasp-
ing the situation as it really is, perceiving the information, the data,
with as little prejudice as possible. The second is rejecting the non-
essentials—the junk, the garbage, the information that doesn't bear
on the issue—and holding in mind only the essentials and the
connections between them. The third dimension is seeing with what
Clausewitz called "the mental eye." Now what is "seen" is no longer
objective. You are the only actor in a drama going on inside your

head. You are distilling a decision from all your synthesized knowledge, experience, instincts, mental courage, and more. Ordinarily the right decision comes to you in a flash of intuitive insight. The details will have to be fleshed out, but generally you've got the answer and know what has to be done. It is always accompanied by a powerful surge of confidence, a sense of being absolutely right no matter what anyone else says. And usually it's simple and uncomplicated. To increase your *coup d'œil*:

Train yourself to see as objectively as possible. "Perceptual objectivity" is the skill of seeing not what you want to see but what is there. Many people are so intent on finding a certain answer that they completely overlook the unusual, the exception, the odd thing. Every scrap of information helps, no matter how random or apparently unimportant and disconnected. In his study *Presidential Power*, presidential adviser Richard Neustadt stated that it is the odds and ends of detail that pieced together in his mind helps the president to make decisions and not the "bland amalgams" of summaries and surveys.

Look especially hard for the critical piece of information. The person with a highly developed *coup d'œil* doesn't need a lot of information, just the critical piece. The problem for corporate managers today is not insufficient information but too much information to make sense of. Each year a million and a quarter new articles appear, and the number of technical and scientific publications in the United States has doubled approximately *every* twenty years since 1800.

The ready availability of information hasn't eased the problems of strategic decision making as expected but ironically has made it more difficult or even impossible. When the human mind can't handle or make sense of the volume of information assaulting it, it responds in certain predictable ways. One of them is to escape from making any decision at all.

Develop judgment through practice. Machiavelli advised his Prince, even when his domain was at peace, to act as if it was at war.

And, too, he said, the Prince ought to study the actions of eminent men to see how they acted in warfare and what brought them victory.

Sweat and fret. Your *coup d'œil* is a product of total absorption in the problem to be solved—total absorption. Alfred Von Schlieffen was one of the greatest of all military planners. His absorption was always total. His dying words were "Keep my right flank strong!" On one occasion while traveling by train, his assistant commented on the beauty of the mountain valley they were passing. Von Schlieffen, his mind always on war, replied only, "An insignificant obstacle."

"The military leader," said Napoleon, "must be capable of giving intense, extended and indefatigable consideration to a single group of objects." Napoleon could. His contemporary Caulaincourt wrote:

> He spared neither pain, care nor trouble to arrive at his end, and this applied as much to little things as to great. . . . He always applied all his means, all his faculties, all his attention to the action or discussion of the moment. Into everything he put passion. Hence the enormous advantage he had over his adversaries, for few people are entirely absorbed by one thought or one action at one moment.

CHAPTER THREE

Maintaining Your Objective; Adjusting Your Plan

An army should have but a single line of operations
which it should carefully preserve, and should abandon
only when compelled by imperious circumstances.

Napoleon

THE BATTLE OF ARBELA

In 331 B.C. Alexander the Great pushed northeast from Egypt at the head of an army of veteran Macedonian troops to face the Persian king Darius in battle. He came in sight of Darius's huge army on September 30, about twenty miles from the city of Arbela. Some accounts put Alexander's numbers at forty-seven thousand and Darius's at one million. These figures are exaggerated, and the precise numbers aren't known. But whatever the exact figures, we know that Alexander was greatly outnumbered. We also know that the confidence of the Macedonians in their leader was supreme and morale was high.

In addition to size, Darius held another apparent advantage: two hundred chariots bearing an innovation—wheels that had been outfitted with scythes. To increase the maneuverability of the chariots in battle, Darius had leveled and smoothed the ground in front of his encampment.

On the morning of October 1 Darius drew up his troops in one immense line. He intended to exploit his main advantage, size, by attacking across the leveled plane, hitting Alexander head-on, using the scythe-bearing chariots to cut through the thick Macedonian phalanx of heavy infantry sixteen ranks deep, and then to release his cavalry in a mass charge to rupture Alexander's center.

Realizing that either to wait or to attack directly across the leveled space was to play into Darius's hands, Alexander, outfitted in a brilliant-colored uniform, immediately maneuvered his right wing (which he always personally commanded) in an oblique, half-right direction not against the Persian center but their left flank.

Immediately surmising what Alexander had in mind, Darius sent the heavy chariots and cavalry of his left wing to swing across his enemy's path. Alexander's light troops defeated them, and for a few important reasons. Alexander's troops were professionals in the best sense of the word. They knew exactly what they were there to accomplish and what they had to do to accomplish it. They knew that their training over the years had prepared them to handle any situation that might suddenly crop up. They never doubted that they possessed every skill they needed and were better at their jobs than the men facing them. They weren't fools; all they had to do was look around them to see they were ridiculously outnumbered, yet they *expected* to win. They had the cool self-confidence of the experienced professional in any field who has been through it all before and because of past successes is fearless and doesn't crack under pressure. Moreover, they were led by a charismatic leader, and a crafty one. Alexander had assigned bowmen to wound the horses drawing the scythe chariots and light armed troops to run alongside them and snatch the reins from the drivers—so much for that innovation.

Darius then made the mistake that cost him the battle and his empire. He ordered the main body of troops on his left wing to attack Alexander's right flank and in so doing opened a huge gap in his line.

Seeing the gap, Alexander, applying the principle of concentration (see chapter 4), immediately sent all his available troops plunging into the hole and through it. His cavalry broke in from the flank, and the phalanx advanced and stormed into it from the front. At this critical moment, Darius's nerve faltered. Abandoning his army, the Persian king mounted a horse and fled. With that, the Persian center broke down completely, retreated, and was chased by Alexander's cavalry.

While things were going well for Alexander, his second-in-command, Parmenio, on the other flank, was having problems with a large force of Persian cavalry. Parmenio sent off urgent appeals for help to Alexander, by now far away in pursuit of the retreating Persians.

Alexander's reply came back: "Hold on! The loss of the camp is only a temporary misfortune. Once the main body of the enemy is completely defeated, all minor setbacks will be thereby rectified."

When more messages came to him, and no doubt confident that he had the situation under control, Alexander turned back with part of his cavalry to support Parmenio. Once seeing that the situation had taken a turn for the better and that Parmenio had the Persians in flight, Alexander immediately returned to the main objective—the pursuit and rout of the main body of the enemy.

The Battle of Arbela was a stirring victory for an underdog (paralleling General Motors surpassing once-powerful Ford in the auto industry. In 1921 Ford was number one, making half the cars in the United States; GM was second but produced only 10 percent of the American cars. By spring 1923, GM, under Alfred P. Sloan, had won its battle with Ford). Arbela was also a monumental moment in world history. When Alexander broke through that hole in the line, he sounded the death knell of the "invincible" Persian Empire—once the ruler of all it surveyed—and marked the beginning of the flowering of Western civilization. Our lives even today would be very different had Alexander lost.

Alexander's stunning victory at the Battle of Arbela illustrates any number of truths of war—the importance of morale, training, seizing opportunity the moment it appears, outthinking the enemy leader, rapid concentration, firepower, and gutsy leadership. But no

truth is illustrated more clearly by this battle than the PRINCIPLE OF
THE MAINTENANCE OF THE OBJECTIVE.

Alexander's objective was as clear as fine glass to him. It wasn't to
solve problems back in the rear; it was to rout the main body of the
Persian army in front of him. Knowing full well that any problems
behind him would be solved by destroying the force to his front, he
pressed the attack forward.

The principle of the maintenance of the objective is different from
the other principles of war. The others are all concerned with "how"
you will win; this one alone deals with "what" you intend to win. Its
major lesson is very direct, very simple: "Pursue one great decisive
aim with force and determination"; "Don't wander off on tangents";
"First things first"; and "Even if you see sitting ducks, but they are
not the important ducks, don't chase them." For example, during
World War II's Battle of El Alamein British general Montgomery
had a large pocket of Germans surrounded. They were ready to be
taken, but Montgomery went around them instead, continuing the
pursuit of his main objective, the German panzer divisions.

"War means fighting for definite results," said French general
Ferdinand Foch. "I am not waging war for the sake of waging war."
The principle of the maintenance of the objective is fighting your
wars for your definite results. It's one of the main forms of business
warfare's competition within the competition. All other things being
more or less equal, the side that's clearest about its objectives has a
better chance of winning than the competitor that's hazy about what
specifically it's trying to accomplish.

How important is the lesson to maintain your objective? It was the
essence of the Duke of Marlborough's approach to war, and he *never*
lost. To penetrate the enemy's center was all he ever wanted. No
matter how successful he was on his two wings, he never deviated
from the center, even if it meant restraining his officers to his right
and left.

Armies that pursue one objective or at most a small number of
major objectives tend to win; corporations that do the same also tend

to win. McKinsey & Company consultants compared the management practices of high-performing companies with comparable firms whose performance had been "not outstanding." *Every* high performer maintained a small number of clear overall objectives such as "Our company is built around customer service," "Pioneers in banking," etc. Ordinarily the poorer performers didn't have such clearly defined statements of overall purpose.

The secret is to know what your dominating objective is and to broadcast it throughout all ranks so that everyone knows. In his book *Henderson on Corporate Strategy*, Bruce D. Henderson also points out the advantages that a company gains simply by being clear about what it's out to accomplish. "Strategy concepts need to be explicit in order to be executed in a coordinated fashion in complex organizations. . . . Those managers who can conceptualize their strategy and make it explicit in terms of a system of competition will dominate their businesses in the future."

A particularly effective corporation distills a small handful of key guiding ideas or central themes that sum up what the firm stands for and is trying to accomplish. Those ideas/themes are then translated into strategies that cut across all units of the company. Those strategies and those ideas and themes, in turn, form a background from which individual tactical decisions and actions can be carried out. So, during day-to-day business operations the company's main ideas/themes provide a broad frame of reference for a manager's or salesman's or staff person's decisions and choices. These ideas/themes range from (and these are the actual, stated themes of a handful of American corporations) "Maintain product excellence," "Grow as fast as the industry," "Be technically superior to competitors," to "Increase dividends faster than inflation," "Lead in manufacturing," and "Acquire new lines of small business with growth potential."

I was asked by a major business-education corporation to recommend new services and product packages that it might offer. I developed a list of a number of them, each of which I could prove would make money and expand the company quickly. Some of the ideas were simply fantastic, and I was excited about their profit potential.

However, during the meeting with seven of the company's marketing executives the ideas weren't generating much enthusiasm. Then it struck me. "Growing slowly and cautiously" was the single most dominating theme of this company and thus of the people I was making the presentation to. That's why they were all squirming. My ideas would make the company a lot of money, but that wasn't so important. What was *more* important to them was that my ideas would force them to expand fast. *That* they were not philosophically prepared to do. From that experience I learned to begin my consulting with a careful consideration of the one, two, or three main themes that the client company values above all others. And I have found in every instance that if the company is a competitively successful one, those few themes are always there and are always very simple. In the same way, after much deliberation, the French general staff in World War I based its approach to battle on but two simple themes: élan (position, offensive, attack, energy, and action) and sureté (preparation, staff resources, and vivid awareness of situational realities). As we'll see in chapter 6, the entire strategic framework of the guerrilla—the single best form of warfare ever devised—is so simple that it can be stated in just twenty words.

IT'S EASY TO VIOLATE THIS PRINCIPLE

If you're thinking that maintaining your objective is so self-evident that you might as well just skip it, think again. Even skilled generals and corporate managers sometimes violate this "obvious" principle.

For example, during the Russian campaigns of World War II, Hitler brought catastrophe to his army by diverting it from the main objective, the annihilation of the Russian army, and sending it after the far less significant oil fields of the Caucasus.

In the middle nineteenth century, Jay Gould's Erie Railroad fought with Commodore Cornelius Vanderbilt's New York Central. In one battle, Vanderbilt lowered the rate for hauling steers from Buffalo to New York from $125 to $100 per carload. Gould counterattacked, lowering Erie's rate to $75. Vanderbilt then went to $50, then Gould to $25. Vanderbilt filled every car when he lowered his rate to a rock-bottom one dollar. Unfortunately for the Commodore, the cattle he was carrying were Gould's. Gould had bought up every steer he could find in Buffalo and shipped them on Vanderbilt's railroad. Vanderbilt had made the mistake of believing that his objective was to carry the most beef on the hoof, while Gould didn't lose sight of the real objective—to make money. Vanderbilt made the mistake of being confused about his objective, illustrating the ancient Chinese military adage "One who is confused in purpose cannot respond to the enemy."

The authors of the book *Business Policy* cite an example of "a large, long-established, diversified and increasingly unprofitable company." The firm's president, after a one-day discussion of the concept of strategy, asked his seven vice-presidents, none of them new to the company, to prepare a one-page summary of the company's business and strategy. They read like descriptions of seven different companies!

THE COMPANY SOUL:
PURPOSES LARGER THAN OBJECTIVES

Those who respect history will deem it beyond argument that belief in a cause is the foundation of the aggressive will in battle.

Col. S. L. A. Marshall, *Men Against Fire*

The Bell & Howell Corporation doesn't make movie cameras anymore. Just about nobody knows that, and I certainly didn't when John Kambanis, B&H vice-president of human resources, asked me if I would address a meeting of the corporation's human-resources-development managers. I asked John for background information on the company, and he sent me a plump stack of reports.

The new B&H newsletter caught my eye—in particular an article written by B&H chairman Donald N. Frey entitled "The Company Soul." The piece begins with a short history of how the company began. Donald Bell and Albert Howell happened to meet one November day in 1906. Bell was a thirty-seven-year-old projectionist in a Chicago movie theater who had trouble keeping the films he ran from jumping their track. After wrestling with the problem, he figured out how to solve it by modifying the projector. He took the projector and the modification specs to a small machine shop on that November day, and while he waited for the job to be done, he happened to strike up a conversation with a machinist in his mid-twenties who turned out to be Howell. Bell said, "I've got the idea," Howell said, "I can make it," and the company was born.

That, the article continues, is what Bell & Howell's "Company Soul" is—it's making things that work, that meet specifications, that contribute to productivity and are useful as a result.

During my talk I said I was impressed with the company's willingness to talk openly about its "soul." Not many companies would be bold enough to relate everyday business activities like running a machine or checking on an order or calling on a customer to something as grand and as lofty and as spiritual as a "soul." But I think the best companies would. Generals certainly do. Civil War general Sherman wrote in his memoirs, "There is a soul to an army as well as to the individual man, and no general can accomplish the full work of his army unless he commands the soul of his men as well as their bodies and legs."

All significant leaders in any field have had this same insight: to unleash a powerful commitment from a group of people—workers or soldiers or athletes, etc.—those people must be supplied with meanings that are larger than the individual. "It is both the noblest and safest thing for a great army to be visibly animated by one spirit," said Archidamus II of Sparta twenty-three hundred years ago. That spiritual animation is what Bell & Howell was attempting to reach through discussing its "soul." It's quite proper to speak of something internal and spiritual like the soul in a corporate context where enthusiasm for the job can make such a difference. The word enthusiasm comes from the Greek *enthousiasmos*, meaning "the presence of a god within."

In their book *The Art of Japanese Management*, Richard Pascale and Anthony Athos state that "great companies make meaning" through "superordinate goals" such as serving the country or fostering the general welfare of society. Japanese companies are certainly not the only businesses making meaning. Delta Airlines considers its secret of success to be the high value it places on service to the customer and a family feeling within the company. At Delta it's said, "You don't just join a company, you join an objective."

What we're talking about are not really objectives at all. We're addressing something that's *higher* and at the same time *deeper* than a specific objective—A SENSE OF PURPOSE. It's the relentless commitment to a purpose that often distinguishes successful individuals and corporations from the also-rans. Purposes can capture the imagination of an entire army and a whole work force and ignite collective human energy. It's widely recognized that only about 25 percent of the American work force finds fulfillment at work. One of the biggest contributors to the dissatisfaction among the other 75 percent is a sense of meaningless—the "What does it matter, anyway?" syndrome. Workers who don't commit to a purpose disconnect. They put in time; they count hours. The word "objective" is used so often in business that it's lost much of its meaning—"Oh, *another* objective, big deal." Objectives only engage a person's brain; a purpose—if it's the right purpose—grabs hold of a person's spirit. And it's from the spirit that generals and managers, soldiers and workers, get their energy and drive.

Any corporate executive attempting to mold a work force into an organized group of people committed to a common sense of purpose could do no better than to study the life of Genghis Khan (1162–1227), grandfather of Kublai Khan. Genghis, the supreme emperor, devoted long years to fashioning a disconnected conglomeration of warring tribes of superstitious, jealous, ignorant thieves and bandits into the finest, best-organized, and most efficient army the world would see for the next six hundred years. Single-handedly, through the power of his personal vision of what could be, he created the powerful Mongol Empire. During thirty-seven years of expansion through conquest, the Mongols added much of the empires of China, Tibet, and Persia, along with southern and central Russia and parts of India and Afghanistan, to their empire. The achievements of Genghis Khan are comparable to an owner of a small computer corporation successfully organizing hundreds of insignificant computer firms into a cooperating force that together outcompetes IBM, Hewlett-Packard, Digital, Xerox, etc. If it weren't for the cruelty of the man, Genghis would be considered one of the greatest leaders of history. Why did men follow Genghis Khan? For two reasons: because he convinced them of his "divine purpose" and because they knew if they followed him he would make rich men of them.

Nietzsche wrote, "He who has a why to live for can bear almost any how." If there's a purpose, there's a will; and where there's a will, there's a way. Throughout history people have demonstrated that they can overcome just about any hardship and any setback and still keep on fighting if they are powered by a purpose. A leading magazine has called advertising company president Barbara Proctor one of the most influential women in the United States. She and I were being interviewed together on a television show, and when I finished talking about the importance of purpose, she added that she had learned the same thing from her own experiences. "The person can't be defeated who's willing to get up one more time." A purpose is not the same thing as the objective but is closely associated with it. Your company purpose provides the fuel—the commitment and energy—that employees pour into the objective.

Someone once said, "It is ideas that inspire courage." Let's shorten that to "it is ideas that inspire" period. Ideas make up corporate souls; ideas define great purposes. "To be the best, the absolute tops,"—or "to maintain product excellence," "to be the winner in each of our businesses," or "to compete by offering unique products," etc.—is the idea inspiring some corporations.

The best group of workers I ever managed was a team of consultants who endured long night hours and weekend work, applying themselves without complaint, rallying each other when one was down—for the simple reason that they were the finest employees in all the company and they were proud. The company soul of Piggly Wiggly Carolina, the Charleston, South Carolina, grocery and distributing business, is also pride. Employees wear lapel buttons declaring it: "I'm Piggly Wiggly PROUD." To show their gratitude to their bosses, the employees got together and gave the company a new refrigerated semitractor trailer. They borrowed the idea from Delta Airlines employees, who gave Delta a thirty-million-dollar Boeing 767 in 1982.

Frederick the Great said a battle is lost less through the loss of men than by discouragement. The First National Bank of Chicago found him to be right when, as a result of a previous chairman's efforts to clamp down on excesses, morale plummeted. The chairman was replaced by Barry Sullivan, whose first task was to get the bank's thousands of employees excited again and working together. One method he used was to give them a purpose. He involved everyone in expanding the bank's credit-card business and increasing its portfolio of IRAs.

The idea that animates Matrix Corporation of Orangeburg, New York—makers of image recorders—is social commitment. Its president, Franklin G. Bishop, hires one-third of his employees through social agencies. His work force is made up of Southeast Asian refugees, Soviet immigrants, battered women, unskilled youth, and other recruits from government-funded job-placement agencies. The company not only provides a service to the needy but produces a product so good that at least one competitor, GE, buys Matrix's products over its own similar model.

The animating idea at one Honeywell plant producing printed circuits was increased efficiency. Workers were told that when they reached their week's quota, they could start their weekend early. Plant managers thought quotas would be reached on Friday about two o'clock. In fact, it happened at noon on Thursday.

The purpose animating small but fast-growing Quill Corporation, wholesale business stationery supplier, according to Jack Miller, Quill's president, is each employee "being the best we are capable of being so we can make Quill the best it is capable of being" and being nice in the process.

Apparently one of my company's bills hadn't been paid. I was shown the request for payment Quill sent because it struck the person doing the check writing as being so pleasant. The notice contained four sentences, the first letting us know that a bill hadn't been paid; the next three apologizing for bothering us. What the notice had that most business letters don't have was a personality. I asked around about Quill, and I was told by one person after another, "They're always so helpful and so cordial," "They go out of their way," "They always get right back to you," "They take time with you."

Any company that's so highly thought of I've got to see, I thought. Taking my six-year-old son along, I went to pay the bill myself. People were right. I had never seen such service and such niceness from any company. For example, after paying the bill, I asked if there was a soft-drink machine close by. It was a miserably hot and humid day, and my son was thirsty. A supervisor told me that unfortunately the nearest machine was far away on the other end of the facility. *No drink*, I was thinking when she said, "So just have a seat and I'll get it for you." I gave her the coins, sat down, and a few minutes later she was back with the drink. Now no doubt she had lost time that could have been spent on something else. But in the long run she had gained far more for Quill. A "nice" company like that you return to—and tell other people about.

THE STRATEGIC AND TACTICAL CHAIN

... the art of war in its highest point of view is policy.
Clausewitz

In the movie comedy *Our Man in Havana*, the blunder-headed chief of the British espionage network, played by Sir Ralph Richardson, asks his assistant where the British spy they are discussing—"their man"—is located. Told, "The Indies," the chief walks to a wall map and peers closely at it.

"The *West* Indies," says the assistant.

The chief says, "Oh, yes," and moves to a map of the Western Hemisphere.

The point is, if there is a commonality of objective from the highest corporate level to the lowest, with operations filling out strategy, everyone should be kept pointed in basically the same direction, looking at the right map and operating in the correct hemisphere.

Q. T. Wiles is head of MiniScribe, a company whose annual sales of personal-computer-disk drives total about $185 million. He keeps everyone pointed in the same direction by asking key questions. At "Q.T.s" once-a-quarter group meeting, his sixty or so executives are each required to fill out answers to the following: "Q.T., this is my charter"; "Q.T., these are my five most important tasks"; "Q.T., these are my operating principles"; "Q.T., these are my shortfalls."

Strategy and Tactics

In the strict sense, military, or pure, strategy means "the art of the general," from the Greek word *strategos*. The primary logic of corporate strategy—"understand the situation and adapt to it"—is based on military strategy. For whatever reason, no one knows the precise derivation of the word "tactics."

Strategy and tactics are different in scope and focus. Tactics are small in scope, involving small movements of forces, narrow spaces, and short time spans. Strategy is large—large bodies of forces, wide spaces, and broad expanses of time.

Strategy might not be only "the art of making war upon the map," but it certainly involves mapping out the broader scheme of things, the "big picture."

Theoretically, strategy precedes contact; tactics begins with it. Stanford's Harold Leavitt sees managing as an interaction among three activities: pathfinding, decision making, and implementation. Much of the growing discontent with corporate strategic planning is caused by a false belief that strategy formulation and implementation are distinct processes, that the pathfinding and decision-making activities are separate from the implementation activity. In reality, there is a feedback loop swirling between strategy and tactics whereby your strategic decisions of today are influenced by the effectiveness of your tactics of yesterday. For example, if the one grand organizing idea, the core concept of your new strategy, has been to offer your product in thirteen different colors and your sales managers feed back to you what their sales reps have fed back to them—that consumers don't want and don't need colors and prefer the good old standard black, anyway, then you might be wise to alter the strategy. It's a serious mistake to think that corporate strategy formulation and its tactical implementation are separate processes or to believe that strategy is more inalterable.

To most, the words "Prussian general" conjure up images of stuffy mustached old codgers who valued logic and precision above everything. Not so in the case of Prussian field marshal Helmuth von Moltke (1800–91). Maybe it's because he wasn't Prussian by birth. But in any event, had he somehow been able to hear that in the late 1980s increasing numbers of corporations would be eliminating "strategic planners" and asking line managers to plot strategy, he would have said, "Fine idea." To Moltke, one of the truly great strategists, real war wasn't at all as neat and tidy as it appeared to be from a war room. It was messy, disorganized shifting of conditions in which victory was often attained *in spite of* preplanned strategy and because of tactical genius. Moltke's advice? "In the case of tactical victory," says Moltke, "strategy submits." In other words, if your staff comes up with something that's effective, don't reject it merely because it doesn't fit in with your strategy. Change the damned strategy! That's what Montgomery Ward did, showing the flexibility to change even after being in business for half a century. Wards opened in 1872. For the next fifty-plus years its sole business was mail-order selling. In 1926 Wards opened a few exhibits around the country so that customers could actually see what they were buying. When one customer insisted on not waiting for mail delivery but buying a floor model, a local Ward's manager spontaneously made the decision to sell it to him. Word got out quickly, and hundreds of buyers flooded the exhibit and bought everything. Changing its entire corporate strategy as a result of this tactical success, Wards opened retail stores at breakneck speed—more than five hundred of them in just three years.

Strategy and tactics also differ with regard to the "knowability" factor. Strategists don't begin with the difficult to know but with the impossible to know, the unknowable. No corporate strategy maker can predict the events that could arise to shape the company's future course or how those factors will interrelate, not even the strategist aided by the most sophisticated computer capabilities. All he or she can do is to forecast what's likely to occur. Tactics deals with the more knowable. With tactics your choices are fewer, your data are clearer, and the look of the land is more distinct. That's why there is always an abstract, rarefied quality to strategy and a touch of vivid reality to tactics, like dirt under your fingernails.

However removed from the front lines of selling, relating to customers, etc., corporate strategy formation is, its sole purpose is to get something done—the "something" being the achievement of the company's objectives. And no matter how uninterestingly complex the corporate strategy texts make it seem, formulating strategy is fundamentally coming up with a good idea for staying in business and prospering. Whenever you find your head swimming with the complexities of strategy, think of General Sherman, certainly the match of most any corporate strategist. He said that when all is said and done, strategy is just "common sense applied to the art of war."

The Four Tiers of Strategy and Tactics

Since the early twentieth century military theorists have subdivided strategy and tactics into two types each to form the four tiers of warfare:

> Higher (grand) strategy
>
> Strategy
>
> Higher (grand) tactics
>
> Tactics

In military warfare, *higher strategy* is the province of kings, presidents, cabinets, and ministries; in corporations, of boards and chief executive offices. Whoever pulls up a chair to shape a firm's higher strategy must be in a position individually or collectively to direct all of the resources of the company. Higher strategy is the equivalent of policy. Higher strategists, operating at the company's uppermost level, make the fundamental decisions affecting all others, selecting major objectives and priorities: What do we want to achieve? What are our products, our markets and market segments for which our products are designed? How will we reach those markets? What is our end objective? And where will we compete—over what product, territory, or market? Who is our competition? If we do X, what will they do? Then what will we do? What alliances will we have to form to win? What size organization will we need to achieve our objectives, and how will it all be paid for?

Clausewitz discusses the self-doubt that strategists experience and that often keeps them from making tough decisions. The scariest thing about selecting objectives is that out of all the choices available to you only a few should be chosen. The problem facing the strategist is not seeing enough opportunities but far too many. Merrill Lynch, for example, runs America's largest brokerage house. Its objective is to become a major provider of full financial services to corporations and individuals. It has added real estate, banking, and insurance to its businesses, and yet surely it realizes that it has not begun to exploit even a fraction of the opportunities available to it. Sun Tzu writes that the general who wins makes "many calculations in his temple" before the battle is fought; the loser makes few calculations. But out of the "many calculations," the winner must choose a narrow course of action for himself or herself and *everyone* else. That's why Clausewitz observed that "much more strength of will is required to make an important decision in strategy than in tactics."

In hindsight, higher strategy that turns out to have worked seems amazingly simple. Jovan's *products* are fragrances, but what it's selling is *sex*. Since its start in 1968 with the Musk brand, Jovan has never once varied its strategy of associating the use of its products with increased sex appeal. Pontiac's concept is reflected in its slogan "We Build Excitement." It's all simple, direct, and successful. Pontiac began to succeed in the mid-1980s the moment it became crystal clear that its objective was to produce high-performance, sporty, exciting cars. Because of its ability to maintain that objective, its sales climbed quickly—faster than GM as a whole and faster than GM's other four cars—Cadillac, Buick, Oldsmobile, and Chevrolet.

Higher strategy that works gives the impression that it almost *had* to be that way. It's so right that it's easy to forget the thousands of alternatives that were rejected. For example, Roosevelt, Churchill, and Stalin meet and decide that:

- The principal objective of the Allies in the Second World War is the unconditional surrender of Japan and Germany.

- At Stalin's insistence, reducing Germany to submission will be the first order of business.

- While it's tempting to invade the Mediterranean, northwest Europe looks more promising.

- The D day landing will be made somewhere in northwest Europe.

Strategy, the next tier, is concerned with using the resources at the disposal of the corporate manager within his or her theater of operation to reach the objective laid out by the firm's higher strategists. Strategists set out the how, when, and where of operations. They create the firm's plan, define the campaigns that will be undertaken, and indicate where and when the major competitions are to be fought and when they will occur.

Having received the Big Three policy decisions, World War II Allied strategists then:

- Considered three possible areas for the invasion: Normandy, Western Europe, and Pas de Calais.

- Selected Normandy.

- Considered dates, times, tides, the number of men, guns, tanks, ships, and aircraft necessary to successfully make the assault, achieve a local victory, and race through the Rhine to the Ruhr, the vital industrial hub of the Third Reich.

Higher tactics, the third tier, overlaps strategy at strategy's lowest levels. It's where most of the problems are. "Even competent generals have found the transition from planning to execution beyond their abilities." Assigning the right number and types of people to achieve their part of the company's objective, coordinating their efforts, and maintaining a proper reserve to release to exploit success or bolster an attacked front are responsibilities of higher tacticians. Higher tactical decisions are made by managers throughout the company.

Tactics, the fourth tier, controls the actual forces in the field, the competitive combats and maneuvers, achieving, hopefully, the specific objective that this individual combat contributes to the whole, whether it be to take a house, capture prisoners, finish a project on time, or to return to the office with a signed agreement.

The essence of tactics is CONTACT—with the enemy, or the competition, and with the consumers and realities of the marketplace. Tactics may lack the grandeur and sweep of higher strategy, strategy, and higher tactics but need take a backseat to none of them. For it is at the tactical level where, like the rubber meeting the road, the objective meets the sometimes harsh test of doability. Vegetius was a fourth-century Roman whose *Military Institutions of the Romans* was the most important military treatise up to the nineteenth century. In it he writes that the tactical engagement "is a conjecture full of uncertainty and fatal to kingdoms and nations, for in the decision . . . consists the fullness of victory. This eventuality above all others requires the exertion of all the abilities of a general. . . . This is the moment in which his talents, skill and experience show themselves to the fullest extent."

Practically speaking, there should be a harmony in the company's strategic and tactical structure. Whatever the enterprise—whether to automate an office or manufacture and market a product—all parts should be singing the same tune but in different ranges. If you're selling sex or excitement, *everything* should declare it. Each tier above should blend imperceptibly with the one below. The strategists depend on the tacticians and the tacticians on the strategists. The possibility to be avoided is of a finely planned campaign poorly fought or a poor plan that wastes the talents of a fine staff.

One way of increasing the harmony is by making as certain as possible that the strategists "on high" persistently talk with the "soldiers" below:

Soldiers to strategists: "What do you want to accomplish? What do our competitors want?"

Strategists to soldiers: "What can you do? What can the competition do? What do you have? What do they have?"

Soldiers to strategists: "What level of effort are we willing to put into this? What will their level of effort be?"

An Example: Mercury Marine

Mercury Marine Corporation is headquartered in Fond du Lac, Wisconsin. It's a small, old city where the air is fresh, the streets wide, and the people friendly. In the summer most everyone's thoughts turn

to the outdoors, to fishing or boating or waterskiing on Lake Winnebago, Green Lake, or one of the other waterways that almost any
reasonably sized road in the area eventually takes you to. Mercury
Marine provides an example of a company that is clear about its
objective and has designed its tactics to harmonize with its strategy.

For nine years its joint venture partner in the outboard-motor
business was Yamaha, the Japanese giant. In 1983 Yamaha followed
the course of many Japanese companies entering an American industry by becoming the direct competitor of its onetime partner.
Nineteen centuries ago Phaedrus wrote, "An alliance with the powerful is never to be trusted." Mercury Marine learned for itself. The
prospect of competing with Yamaha was intimidating to Mercury's
management, especially since Yamaha had an immediate advantage
of a 30 percent edge in costs. To compete successfully, the company
had to "change its way of doing business and thinking," said Richard
J. Jordan, Mercury's general manager and the second-in-command of
Brunswick Corporation, Mercury's owner.

Mercury has been so successful that with more than three years to
go on its organizational improvement program it had increased production by 46 percent in four years, raised its market share to its
highest point ever, and held Japanese inroads into the American
outboard-motor market to under 10 percent market share. Now
Japanese companies are sending their managers to see how Mercury
does it.

The company's main strategic objective was to increase productivity. To achieve that, it would completely overhaul its sizable plant
and operations.

Ninety million dollars were allocated to purchasing new equipment and training staff through 1992.

Brunswick introduced a decentralized management system that
gave Mercury executives the freedom to implement changes as
necessary.

Mercury's manufacturing system was completely changed. Traditionally it had manufactured by "batching." This consisted of

producing huge batches of one kind of outboard motor for four straight months, then turning out another model for the next four, etc. Mercury eliminated batching and replaced it with a production system that turns out every product model every day. It maintains four lines, each of which can manufacture up to twenty-six models each and every day.

Although originally fearing that the economics of production scale would be lost under the new "every model every day" system, Mercury discovered that this tactical change brought a number of benefits. Expenses of inventory fell—that was expected—but why did quality also improve? Errors declined because machinists, now working on *all* the motors each day, were more familiar with all of them. Making every engine every day also enabled Mercury to be more sensitive to changing market demands.

Another tactical change was in the area of simplifying production processes. Traditional manufacturing plant layout is by specialization —machine work is located in one spot, assembly in another, etc. Often plants look like ant colonies, bodies crisscrossing others in complicated patterns. Mercury organized its production personnel into small units, called "U-lines" because of their shape. Each U-line is responsible for producing and assembling a major part of an outboard motor; each worker is trained to do all the jobs on his or her U-line. Communications improved because each worker is only a few feet from all the others he needs to talk to. Also, Mercury has been able to reduce expensive floor space by more than 30 percent.

To improve delivery of parts, Mercury chose to contract with fewer suppliers. In 1978, 850 suppliers were used; soon Mercury hopes to be doing business with about 100 or 150.

If an employee's department increases productivity, each employee benefits, sharing productivity improvements on a fifty-fifty basis with Mercury.

Realizing that it's not enough to *tell* employees about the Japanese competitor, Mercury has sent about two hundred employees to Japan to see for themselves.

Mercury's strategic objective is increased productivity. *Everything* is designed to achieve that objective, from floor layout up.

SETTING YOUR OBJECTIVE

The most important function of strategy is to serve as the focus of organizational effort, as the object of commitment....
Edmund P. Learned et al., *Business Policy*

Market share and return on investment are often used as principal measures of a corporation's competitive strength. If your company is one of the majors, its rank in your industry also matters to you, but if it isn't, you probably couldn't care less about your rank as long as you're showing a profit year after year.

But corporate objectives are limited only by how fertile the brains are of the managers thinking up the objectives. There are financial objectives, market objectives, product objectives, and more. And at times objectives are remote from turning a profit. For example, when General Foods entered the gourmet-foods business in the middle fifties, its objective was not necessarily to show a profit but to improve the company's image with consumers, the financial community, and GF's own employees by offering a line of prestige items.

And of course a corporate objective can be changed with the times. "Whoever desires constant success must change his conduct with the times." Machiavelli wrote it; many companies are doing it. Under founder and genius inventor Edwin Land, Polaroid's aim was clear. "We're not here to make profits," Land once said. "We're here to make innovation." Now under new leadership, Polaroid is after profits, streamlining its $1.3-billion company and venturing into electronic imaging, video equipment, and computer products.

In the mid-sixties sixty-four million passengers left the driving to Greyhound; in 1985 only slightly more than half that number used Greyhound—thirty-three million. The emergence of more than one thousand small competitors, high labor costs, and cheap air travel have conspired to throw Greyhound on hard times. Greyhound is changing with the times by diversifying into business and financial services and by shunting off into charter tours.

The objective your company chooses to pursue in business warfare depends on a bewildering array of factors, including these five important ones:

1. Whether or not your company is a combatant

2. The type of war objectives your firm chooses

3. The means available to your company to wage war

4. Your company's condition relative to the competition

5. Your company's motives for going to war

Each is discussed in the following sections of this chapter.

Combatants and Noncombatants in Business Warfare

The first factor affecting the objective your company sets is what type of combatant it will be or if it will be one at all.

Offensive objectives and strategy and tactics aim to conquer; defensive ones, to defend. Electrolux, the Swedish company, has offense-oriented objectives. Historically, appliances made in one country have not sold well in another. Electrolux, always aggressive, intends to become the first appliance maker to dominate the global market. Its objective is clear. Offensive objectives are further broken down into offensive action to entirely defeat the adversary, to take only one area of its territory, or to shove it back a little.

Neutrals. The majority of corporations are like the Swiss. Comfortable as they are, they prefer neutrality. They try to remain noncombatants if it's possible. They're often in mature industries, but they needn't be. Any company of any size in any industry that's generally content with its status and doesn't want to make waves is a neutral. They need not study the rules of offensive action, because they rarely seize the offensive. They set defensive objectives.

The Fittest. In almost every industry one company or two has the largest share of the market and dominates the industry. These are the fittest, as in the survival of the . . . There is Texas Air Corporation in airlines, Coca-Cola in beverages, Du Pont in chemicals, American Home Products in drugs, General Electric in electrical, Marriott in lodging, IBM in office equipment and computers, and LTV in steel, etc. These fight against other companies in the industry or industries the fittest are in, particularly against . . .

The Challengers. In almost every industry one or more companies are pushing hard for more of a share of that industry. These are active aggressors—United, Pepsi, Dow, Merck, Westinghouse, Hilton, Hewlett-Packard, Bethlehem, etc., and their aggressive objectives often include a direct reference to the fittest. Their objectives make no secret of the company they're competing against.

Guerrillas. These are smaller companies, somewhat removed from the actions of the industry's giants. They survive, as all guerrillas do, by gaining and keeping the loyalty of the people, in this case the people comprising the ordinarily small, narrow market niche that these companies serve. Out in the hinterlands (always figuratively; sometimes literally), the guerrillas may stay small or grow, depending on how successfully they fight. If successful, the guerrilla corporation, exactly like its military counterpart, will make the transition from guerrilla fighting to conventional warfare. I've eaten at the first McDonald's opened by the late Ray Kroc. It's easy to forget that from that one little drive-in restaurant nine thousand others would evolve and that at one time not so long ago McDonald's was nothing but a small guerrilla operation in the hinterlands of Des Plaines, Illinois. In the last stage of its special kind of warfare, the guerrilla will leave the remote provinces and march its army straight to the capitol (major markets), where it will do battle with the neutrals, the fittest, the challengers, and the . . .

Retreators. These companies are bailing out, liquidating rather than continuing the fight or discontinuing marginal operations.

Hybrids. We can talk as though any one company is either one type or another. In fact, companies can be hybrids. In very large companies one of its units may be the fittest, while another may well be a guerrilla and another a neutral or a retreator. Diversified companies survive by regulating the combatant status of their assorted businesses. They channel internal investment from some businesses to others, just as a government invests more resources in one theater of war than another, depending on its priorities. Many diversified companies classify their business units on the basis of building, holding, or harvesting. Building, the strategy of challengers, is an offensive objective. Holding, the strategy of staying neutral—the most common corporate objective—is defensive. Harvesting is a retreat strategy of allowing market share to decline so that cash can be used elsewhere. It's usually forced on the company and is positive if by yielding that ground a more valuable objective is gained. Most every general has sacrificed cities, terrain, and forces to gain a greater objective later on.

The Two Types of War Objectives

Battles are won by slaughter and maneuver. The greater the general, the more he contributes in maneuver, the less he demands in slaughter.

Winston Churchill

The second factor bearing on the objectives you'll pursue in business warfare relates to which of the two broad strategic types you consider preferable.

"How can we best win this war or this campaign?" is the fundamental question all great military commanders have asked themselves. Corporate managers and owners ask themselves the same question. And like their military counterparts, they discover that to that one question there are but two answers: (1) by trying to annihilate the opponent or (2) by outmaneuvering him. All strategy can be divided into those two basic forms: the strategy of annihilation and the strategy of maneuver. You're already familiar with them. As a matter of fact, whenever you devise your own moves vis-à-vis a competitor, these two options are in the back of your mind.

The strategy of annihilation is also called the strategy of total war, or absolute war. Total, or absolute, warriors of history have aimed at the complete destruction of the opponent and/or his unconditional surrender. Theirs is the strategy of the fight to the finish, no holds barred, no resources spared, until, as with the *torero*, the final sword is put into the bull and the competition no longer exists.

The means annihilators use to achieve complete destruction of the enemy is the *great battle*. The strategy of annihilation is sometimes called "one pole" strategy, referring to the one "pole" or option on which it's based—the great battle. Clausewitz, the major theorist of war of annihilation, was also the advocate of the great battle. He wrote, "There is . . . nothing in War which can be put in comparison with the great battle in point of importance, *and the acme of strategic ability is displayed in the provision of means for this great event, in the skillful determination of place and time, and direction of troops, and in the good use made of success.*" "The battle" said Clausewitz, is "war concentrated."

The ultimate focus of annihilators—and their perfect strategy—is to arrange for the great battle or a quick run of battles that brings the opponent to ground and destroys him. The bygone days of American business had its total war annihilation commanders. John D. Rockefeller of Standard Oil fought business war so total that in seeking the complete annihilation of his competitors, he even did battle with and destroyed his own brother's company.

If the war of annihilation were the one and only type, then only those nations and companies with the means to destroy the opponent completely could ever wage war. Only the powerful would make war. Your own experience belies that view. Every day in every industry less powerful companies are actively and successfully warring against the giants as well as against other large, medium, and small companies.

The closest military parallel to current business warfare is not the world wars of this century with their total war objectives but eighteenth-century warfare up to 1780. In that century, warring dynasties fought to achieve limited objectives very much like a version of a present-day corporation's market-share increase—not to capture the opponents' whole country but a province, or to adjust a boundary; not to fight a battle but to avoid head-on conflict by maneuvering. If one side maneuvered with sufficient skill to place the opponent in a checkmate position from which, if he fought, he would surely lose, no actual pitched battle, always expensive in terms of men and material, was even necessary. The same kind of ingenuity in maneuver and position that we witness time and again in today's business warfare was prized by warriors in the eighteenth century, more highly prized than brutal combat.

Sometimes called limited-aim warfare, this form of combat is also referred to as "two-pole" strategy. It's called "two-pole" because it offers two polar-opposite options from which the commander or manager may choose. One pole, one option, and the main one is maneuver; but there is another choice, battle. Battle is the second pole. It's between maneuver and battle that "the decisions of the general move" if he uses two-pole strategy.

The epitome of the two-pole strategist is the guerrilla. Fighting "the war of the flea," the guerrilla maneuvers incessantly, then goes to battle only at the most opportune moment. The combination of maneuver and battle is so potent that if the mix of those two poles is correct, the guerrilla is virtually impossible to beat.

It's important to remember that under this "two-pole" system you may engage in a great battle but also to recall that you don't have to. Battle isn't the single method of your strategy. It's one means of reaching your strategic ends but not the only means. For example, Maurice de Saxe, marshal general of the armies of France in the mid-1700s, great in both theory and actual application at the head of troops, believed that battles weren't necessary at all! He wrote, "I do not favor pitched battles . . . and I am convinced that a skillful general could make war all his life without being forced into one. . . .

Frequent small engagements will dissipate the enemy until he is forced to hide from you."

After the decline of the Roman Empire, the commander Belisarius, another two-pole strategist, almost single-handedly revived Roman dominion for a time. Little known by the general public but among the first ranks of warriors, Belisarius won consistently under the most difficult of circumstances—far from home, at the head of a small army at times made up of mere raw recruits. While Belisarius at times waged battles, his principal strategy was to avoid them. This military master conquered by avoiding battle.

The strategy of annihilation and that of maneuver have each had their great exponents. Alexander the Great, Caesar, Napoleon, and Clausewitz were strategists of annihilation. Pericles, Belisarius, Gustavus Adolphus, and Frederick the Great were advocates of maneuver.

If your objective is total war, you should arrange great battles with the competition, matching their model X with your model XX; if they become active in the East, you should meet them there. If they lower price, you should lower it more, etc.

If your objective is two pole, you may at times battle but usually maneuver. Your credo will be something like this:

- Our strategy is not based on defeating our competitor's main strength in Clausewitzian sense but on the notion that it's what we do with what we've got that wins. Adapting our means to achieve our objectives—that's our secret of competitive superiority.

- The end is not necessarily to battle competitors directly. There are other paths to profitability. We can even win by avoiding direct confrontations almost completely.

- We should maneuver toward opportunity along the lines of our strength and competitor's weakness.

- Maneuver is preferable, but if impossible, we will do battle.

- The only time for battles is when we're strong enough; until then we will wait.

TRW is the six-billion-dollar conglomerate in defense, electronics, and industrial products. It provides a case of a major corporation that has discovered that maneuver is usually preferable to battle. When TRW tried to expand its credit-data operation to cover businesses as well as individuals, it ran up against Dun & Bradstreet, which held 75 percent of that market. Learning that success comes hard when your competition is that powerful in a market, TRW took a beating until it pulled back. It now targets promising niches in title and appraisal-data services.

Nowadays, almost all forms of conflict and competition are limited to one degree or another. Nations avoid total war, football teams try not to run up the score, and powerful corporations don't ostensibly set out to drive the competition to Chapter 11. Most business warfare is based on the two-pole strategy of constant maneuver and sometimes battle. Completely unrestrained corporate warfare, victory at all costs, would eventually destroy all companies involved except one. That one would be the king, as in the children's game "king of the hill." But this king would have an empty treasury. In actuality, most business warfare is tempered by self-restraint by all sides—although some restrain themselves far more than others. The tacit understanding in most business warfare is "Never press the competition beyond a certain reasonable limit." Even corporate battlers never throw *everything* into one competitive battle.

But neither is business warfare a completely gentlemanly or gentlewomanly thing. There are corporations fighting limited war "as if" it were total war of annihilation. Antitrust laws are designed to curtail business-war strategies of annihilation aimed at wiping out the competition. Nevertheless, there are entire industries in which, while perhaps not intended to kill the competition, the combat has the same ferocity and the great-battle orientation of any war ever fought by Alexander the Great or Napoleon.

Air war (the competition of airlines), beer war, chip war (the U.S. semiconductor industry), car war, Coke/Pepsi war, health-care war, the large Japanese/American war, movie-studio war, computer war, and software war—these are just a few examples of areas of corporate competition where the warring is hot and where few holds are barred.

The Means Available to Your Company to Wage War

This third factor affecting your objectives concerns the means at your disposal with which to compete.

To fight and win wars of annihilation, a company would have to be willing to compete all out and possess material superiority and advantages in morale and spirit, be staffed with highly aggressive managers, and be willing to incur extreme hazards. To fight and win a two-pole war, you need all of the above, not generally, but along the lines of your concentration.

In business and warfare alike it's difficult to raise a small power to a great one. In industry after industry the winners' list of dominant companies is pretty much the same from year to year. Yet history tells us that it's possible to upset the balance. Alexander the Great, Gustavus Adolphus, and Charles XII of Sweden each used small but disciplined armies to rise from minor to major powers, eventually overthrowing anything that opposed them.

The secret of each of these eminent warriors was never to pursue an objective that was beyond their ability to accomplish. Each looked hard and long at their means, then adjusted their ends to them. "We must consult our means rather than our wishes," wrote our own George Washington. In determining your objective, no sense is more important than the cold, clearheaded sense of what is doable and what isn't, given your side's resources and its competence. Never bite off more than you chew; never demand more than your people and financial and material resources have the capability to achieve. At Austerlitz (1805), Napoleon's plan was based on his making a two-pronged assault against the Austro-Russian lines, but instead he attacked obliquely on one line when he saw that the troops he had intended to use to hit from the south were too exhausted to fight hard. He didn't panic, and he didn't give up his objective. He just reached it differently. And it's important to remember that he still pulled out a victory—a glorious one. Austerlitz was Napoleon's masterpiece.

At times the best thing a strategist or tactician can do is *wait*—for a shakeout in the competition, a technological breakthrough, the completion of a training program, a trade agreement, a management change (on your side *or* theirs), or a lucky break. One of Montgomery's secrets of success at El Alamein was his ability to wait to attack until everything was ready.

Your Condition Compared to the Competitor's

And as water shapes its flow in accordance with the ground, so an army manages its victory in accordance with the situation of the enemy. And as water has no constant form, there are in war no constant conditions.

Sun Tzu

The fourth factor bearing on your objective asks you to compare yourself with the competition. Victory is always worked out in relation to your foe. Unless you're extraordinarily powerful, the objectives you establish are dependent on the competition's objectives, their strengths, their weaknesses.

Whole long shelves could be filled with fat books on competitive analysis. It's doubtful that in total they would have more to say than Sun Tzu. He claimed that any commander will be able to forecast victory or defeat by answering seven short questions.

- Which of the two commanders exerts the greater positive influence over morale?

- Which of the two commanders has the most ability?

- Which side has the advantages of terrain and weather? (Even the market has its "weather," including changes in consumer preferences and tastes, and for terrain corporations have market segments.)

- On which side are instructions and regulations better carried out?

- Which army is stronger?

- Which has the better and more thoroughly trained officers and men?

- Which side is more enlightened in administering rewards and punishments?

The side will win, adds Sun Tzu:

- Which knows when to fight and when not to

- Which knows how to handle both superior and inferior forces (is skilled at both strategy and tactics)

- Whose army is animated by the same spirit throughout all ranks

- Which, prepared itself, waits to take the enemy unprepared

- Which has the capacity to win and whose managers are not interfered with by the "sovereign"—the highest levels of the company

Your Motives for Going to War

This fifth factor affecting the objectives you pursue concerns your motivation for warring.

Your objectives and strategy will differ according to your reasons for engaging in war. Companies go to war because they expect to profit. It's that simple. The fundamental reason for business warfare is that your competition is the major impediment to your higher profits and growth. Your strategy is an attempt to do something about the fact that the competition succeeds at your expense.

The main types of war are these:

War to reclaim or defend rights.
Your objective when in this type of war is to force the competition to leave the disputed territory. You will take the offensive and try to

occupy the area. The human emotion fueling wars to defend or re-claim rights is revenge. "The flashing sword of vengeance" alone has sustained individuals, armies, and entire nations against a terrible foe: "Remember the Alamo"; "Remember the Maine." Remembering and rectifying a wrong is part and parcel of everyday business, too. Re-venge is one of the principal causes of all wars, but particularly this type. That revenge, or the hope of it, supplies the motivation for business wars, too, is attested to by Coke's reprisals against Pepsi. Long-suffering for years as the challenger, Pepsi, moved against it, Coke, the fittest, suddenly turned on Pepsi with a vengeance.

War to protect and maintain your interests.

If its home territory is attacked, a company thrown on the defensive will usually prefer to fight than to yield without fighting. It may be advantageous to take the offensive when the competition's attack appears imminent.

War of expediency.

Your objective here would be to acquire more territory or to prevent the competition from expanding theirs. If smaller, you will not attack a dangerous rival in the same way you attacked before and lost. For example, after suffering major setbacks in the United States as late as 1982, Volkswagen is attempting a comeback. The territory it has singled out is the high end of the auto market, not the low end it once controlled and lost. It will sell expensive big-engine cars.

War without allies.

In war, an ally is desirable, all other factors being equal. Strategic alliances have been a major feature of warfare since the time one band of cave people made a deal with a second band to attack a third. Corporate strategic alliances are the wave of the future. The director of economics for Olivetti puts it as succinctly as anyone could: "A company's competitive situation no longer depends on itself alone but on the quality of the alliances it's able to form." I purposely cite Olivetti because it's almost exclusively Olivetti's ability to form the right alliances—with AT&T and Xerox in particular—that transformed it almost overnight from a mediocre also-ran to a one-billion-dollar-a-year manufacturer of personal computers that's successfully attacking IBM. Westinghouse/GE, GM/Toyota, GTE/Fujitsu, and Ford/Measurex are just a few examples of relationships that started the tidal-wave trend to strategic alliances.

War of intervention.

In military warfare, when two powers are fighting, the sudden entry of a new and large third party on the side of one or the other is usually decisive. In business warfare, the third doesn't usually join forces with one of the original two but exploits their fighting each other.

War of conquest.

A company can attempt two types of invasion: (1) of a neighboring "state"—a new business or market that is close to its current objective; (2) of a distant point—a new, relatively unfamiliar business or market. The second is more dangerous. Historically, wars of invasion are often prosperous.

Two wars at once.

The Romans warned against undertaking "double wars." Single-product companies abide by the Roman warning, but diversified corporations fight double, triple, and more wars at once. Napoleon did also, voluntarily entering large wars with Spain, England, and Russia at the same time. Two guiding lessons are (1) if possible, avoid double or multiple war unless you're as strong *in total* as the companies in the market niches you're competing for; (2) as a rule, concentrate more emphasis and resources on defeating one before shifting full attention to the others. Circumstances will determine if (2) is possible. But if it is, do what the United States did in World War II. Fighting a double war in Europe and Asia, it focused first on winning the war in Europe before finishing things in Asia.

If Alexander the Great were a corporate consultant today, he would advise the company thinking of attacking the markets of a large firm supported by smaller companies to attack the smaller companies' markets first, the large company later on. Alexander built an empire following this system. Napoleon studied it and followed it to a T. The rule of thumb, obviously, is never spread yourself too thin —try never to be involved in so many far-flung efforts that you cannot outfocus the competition in any or all. Always be able to throw more force into an action at the decisive point than your competitor can or step away.

We have just looked at five important factors affecting the objective you choose to MAINTAIN.

1. *Whether or not your company is a combatant.* You should know if your company is a neutral, or the fittest, a challenger, a guerrilla, a retreator, or a hybrid. The type it is colors every objective, every policy—everything.

2. *The two types of war objectives.* Remember the terms one-pole (battle) strategy and two-pole (mostly maneuver, sometimes battle) strategy, then choose your pole(s). Probably 95 percent of business in the United States is based on two-pole strategy. Like Belisarius, the Roman commander, many corporate managers make big wins for their companies by *avoiding* head-to-head battle with a competitor and skillfully maneuvering around him.

   As for Marshal de Saxe's conviction that a skillful general could make war all his life without fighting a battle—is this possible in business war? A silly-sounding but fine product shows that it is. Tootsie Roll has no competitors—it's a unique product. Everyone knows it; there is no substitute for it. Tootsie Roll never has to battle.

3. *The means available to your company to wage war.* One corporate president told me his company's motto was "Think small and never be original." That's not what "adjust your ends to your means" necessarily stands for, although, ironically, a company can become comfortable thinking small and never being original. You can be relentlessly ambitious and outrageously

aggressive and still follow the axiom "adjust your ends to your means." Alexander did at Arbela even though he was outnumbered by about twenty to one.

If you follow the principles of war and at the same time constantly adjust your ends to your means, there is no reason why you cannot raise your company to a greater power.

4. *Your condition relative to your competitor's.* It's worthwhile from time to time to review the seven simple questions Sun Tzu raises and the list of characteristics of the side that will win.

5. *Your motives for going to war.* You've got to know the type of war—type of competition—you're entering; whether you're going in without allies, to defend your rights, to conquer, etc., before you can fully understand what your objective is. Be able to say, "We're fighting a war of conquest" or "We're about to engage in a double war," etc.

GUIDELINES FOR MAINTAINING YOUR OBJECTIVES AND ADJUSTING YOUR PLANS

The conduct of war resembles the workings of an intricate machine with tremendous friction, so that combinations which are easily planned on paper can be executed only with effort.

Clausewitz, *Principles of War*, 1812

Another form of the competition within the competition of business warfare is over *adaptability*. The competitively superior company will usually prove itself to be better at adjusting to circumstances than the competitor that's not as capable.

Don't Plan to Excess

Mao Tse-tung viewed planning wars and campaigns as giving proper attention to the relationships between the opponent and yourself;

various campaigns and operational stages; those parts that are decisive for the situations as a whole, front and rear; losses and replacements; fighting and resting; concentration and dispersion; attack and defense; advance and retreat; concealment and exposure; the main attack and supplementary attacks; centralized and decentralized command; protracted and quick wars; and positional war and mobile war.

Having said that, we should also add that even after considering each of these issues and more in fine-tooth-comb minuteness, IT'S IMPOSSIBLE IN UNCERTAIN ENVIRONMENTS TO PLAN THE WHOLE CAMPAIGN IN DETAIL AND FUTILE TO TRY TO. Clausewitz wrote that because of the peculiar difficulty of war—that all data are uncertain—every action must be planned as if in a feeble light of twilight that gives things exaggerated dimensions and an unnatural appearance.

The debate that's currently taking place among corporate strategists has long been raised in military circles: is it better to prepare and to stick to a detailed, long-range strategic plan or purposely to allow some vagueness in plans so they can be deviated from when the situation at hand warrants? Judging from history, the latter is preferable. Helmuth von Moltke put it as directly as possible. "No plan," he wrote, "survives contact with the enemy." Strategy itself, he added, "is a system of *ad hoc* expedients; it is more than knowledge, it is the application of knowledge to practical life, the development of an original idea in accordance with continually changing circumstances. It is the art of action under the pressure of the most difficult conditions."

The problem is not with planning. Studies show that companies that plan more are more successful than those in which there is less planning. The problem is overplanning and falling victim to the erroneous notion that plans once made shouldn't be deviated from. Planning to excess and dogmatic enforcement of plans of operation lose wars; elasticity in plans wins. Harold Geneen, called the Michelangelo of management, "in principle" didn't believe in long-range planning. No one is wise enough to see five years into the future and plan for it with any sensible certainty, he said; management has enough to do in planning for one year ahead.

George S. Patton wrote, "Successful generals make plans to fit circumstances, but do not try to create circumstances to fit plans. . . ." "A good plan violently executed NOW is better than a perfect plan next week." And a second-century Roman said simply, "It is a bad plan that cannot be altered."

Shogun Ieyasu Tokugawa, one of the five great generals of Japanese history, made an effort specifically never to give detailed orders. It was always best, he said, to "leave the details to be settled according to circumstances." Alexander, too, preferred giving his subordinates general missions that he changed at will during the ebb and flow of battle, and historian Barbara Tuchman wrote, "Human beings, like plans, prove fallible in the presence of those ingredients that are missing in maneuvers—danger, death and live ammunition."

Corporate executives are often criticized for paying too much attention to individual responses to entrepreneurial opportunities and threats from competitors rather than sticking to the corporate long-range plan. What these men and women are often doing is exactly what Oliver Cromwell and Ulysses S. Grant did. They ALWAYS kept their end objective in mind, prepared a basic (sometimes very basic) plan to achieve it, then deviated from the plan and relied on their initiative when the plan proved inadequate to take advantage of the opportunities revealing themselves right before their eyes. What corporate planners tell us is a "sin," military leaders inform us is a laudable virtue. Erwin Rommel considered routine plans to be fetters from which we must liberate ourselves if they conflict with the facts and possibilities of the moment. Napoleon said, "Plans of campaign must be constantly changed, according to circumstances, the genius of the commander, the quality of the troops, and the terrain."

Winning corporate managers are often like Gustavus Adolphus in his campaign of 1632. Nowhere were his exact plans clearly stated.

He had in mind a final march on Vienna, but from the beginning of the year there was a haphazard look to all his operations. It was broken into little pieces, like a jigsaw puzzle. Yet somehow he had eight full armies on foot and kept them moving toward the objective.

Moltke put into words what managers often believe. He said "the independent will of the enemy" makes it impossible to forecast what war will bring. According to Jack Welch, CEO of General Electric, GE's strategic plans became less useful as the company grew bigger and bigger and as planners spent more hours preparing them and making them even more detailed, embellishing them with graphics and nice covers.

How often we forget that a plan isn't reality but only our hypotheses of probable reality and that any plan must take into account the simple fact that the competition often has the ability and desire to frustrate it.

Clausewitz wrote that planners should define the war's main outline and character according to what probably will happen, but should also recognize that anticipating the future correctly is an immense problem whose mistakes can often be solved by the "flash of genius" of those involved in actual implementation of the plan. Jomini added that the only elements of a campaign that can be planned in advance are the end objective, the general plan, and the first enterprise. What you will do after that depends on what happens during that first enterprise.

The implementation of a plan should be reasonable and fluid. Reasonable in the sense that you should never defy good sense (the Germans did in World War I when they used gas on the Western Front, where the prevailing winds blew it back in their faces) and fluid within limits, with respect to how it is carried out but never with regard to what is to be achieved. There are many paths to the top of the mountain, goes a Chinese saying. Follow one, but get up there.

The fine balance that winning managers and warriors strike consists of clearly establishing their end objective, then following a general line of forward movement in the direction of that objective, but not necessarily the completely straight line that dyed-in-the-wool planners would want.

Take a Line of Operation Offering Alternatives

In the Civil War's Atlanta campaign, General Sherman realized that by sticking to a single geographic objective he was making it easier for the opponent to parry him. After that, he continually worked to place the enemy "on the horns of a dilemma" (his words) by taking a line of advance that kept enemy commanders in doubt as to which city he intended to attack. While preferring one city over another, Sherman was willing to attack either, depending on which objective best presented itself. Undecided as to which he wanted, the Confederates tried to protect both, thus dividing and weakening their forces.

There's a valuable lesson here. In pursuing your decisive objective, make certain you have in mind alternative routes to it. If the first path doesn't work, slip off to another and try it that way. Every plan ought to have "branches," each so well thought out that at least one of them will succeed.

The Koreans are invading American markets. The vanguard of the attack of South Korean industries will be carried out by the huge Korean automaker, the Hyundai group. The Koreans believe that now that Hyundai's Excel car has become the fastest-selling import in American business history, the American market will be softened for other Korean products—TVs, steel, toys, textiles, electronics, personal computers, etc. However, if Korean cars had not been well-received here, another Korean industry—another branch of the Korean business invasion plan—would have taken up the initiative. If that one had failed, another industry would have been emphasized until one had entrenched itself and prepared the way for others. A plan with branches.

In 1984 Tandon Corporation's revenues from the disk drives it made under contract with IBM were over $400 million; the next year the income from the IBM contract fell to $269 million. The same thing happened to a number of IBM suppliers when the computer business nosedived in 1985. As a result, increasingly more of these small suppliers of technology found alternatives to putting all of their eggs in one basket—even if it's IBM's basket. Firms that in the past would have been happy to have an exclusive contract with IBM insisted on nonexclusive arrangements, allowing them to market to IBM's competitors so as to avoid complete dependence on IBM. Some asked for guarantees on sales. Almost all learned the importance of a plan with branches.

It sometimes happens that a company or an army is obliged to change its line of operation in the middle of a campaign. Making the change effectively can lead to great success; handling it poorly can bring disaster. Having your plan with branches in mind beforehand can bring victory. Before the battle of Austerlitz, Napoleon had decided that if he had lost he would pursue his attack through Bohemia on either Passau or Ratisbon.

Improvise

History is full of generals and managers who were adept at planning engagements and bringing their forces to the action but who proved themselves totally incapable of adjusting to circumstances in the field. What has always barred them from full success is their inability to improvise with confidence, exercising control over the situation whatever eventualities arose.

3M frankly admits that it doesn't know what the sales growth of a new product will be. It improvises by first entering the market and then making market forecasts.

Divining the meaning of what is happening and being able to exploit it on the spot is one of the infallible signs of a great commander.

Avoid a Pitched Battle When Your Competitor Is Stronger

I've had the good fortune to have had not one but two mentors in my business career. One was Arthur Greenleigh, founder of Greenleigh Associates, consultants; the other was economist Dr. Louis Ferman. Both of them gave me the same piece of advice when I was young. Had I known Frederick the Great, he would have said exactly what they did: "Never get into a pissing contest with a porcupine!" Knowing when *not* to fight is at least as great a skill as the ability to know when to fight.

In business warfare every decision to sell a product in a particular market is also an act of selecting your competitors. You know that by going in "there," company X will be your chief competitor and that by selling "here," companies Y and Z will do battle. Unless you're a big porcupine, don't select a market already occupied by one.

How Many Objectives?

Texas Instruments' motto that more than two objectives is no objectives at all is a positive expression of the principle of the maintenance of the objective. Hard to follow in a work world in which thirty or thirty-five objectives are vying for your attention? Of course. It's always difficult trying not to be thrown off the track by every little nutshell and mosquito's wing that falls on the rail, as Thoreau put it.

Colgate-Palmolive Inc. is increasing its earnings in the toothpaste war by exploiting just two competitive concepts: innovation and "flash." Innovation helped Colgate to beat Procter & Gamble to the consumer with a pump dispenser and gel toothpaste, enabling it to increase its market share by a creditable 8 percent in six years. Under CEO Reuben Mark, Colgate has replaced its old-fashioned advertisements with more upbeat, zingier ads.

The underlying lesson of the principle of the objective is that the discipline required to reject inessential objectives so as to focus on the really important ones gains victory.

Use Scenarios Rather Than One Forecast

Many corporate plans are based on one specific forecast of the future. "Since X will happen, we'll do Y." Some business environments can be predicted fairly accurately, but when they cannot, alternate scenarios of possible futures are preferable to one forecast. For example, two years before the United States entered World War II, five separate scenarios and sets of plans were developed for possible American participation—code-named Rainbow numbers 1, 2, 3, 4, and 5. When war came, the strategy selected was not Rainbows 1, 2, 3, and 4 but was essentially Rainbow 5. In other words, had we planned around one forecast or two, or even four, we would have been unprepared. It was only because we put in the sweat and toil of working out a number of scenarios that we were able to respond swiftly.

Frederick the Great used his incessantly active mind to develop plans and projects for every eventuality, many of them rough ideas that he scribbled on little bits of paper. Like many managers, he had to write things down before making a decision, and like any number of us, he used comparative categories, on the left side of the sheet listing his reasons for remaining in alliance with France and on the right side recording reasons not to, etc.

The worst possibility in warfare, including business warfare, is to lose. The second worst is to be surprised. Whoever trusts in one or a few versions of what is going to happen in the future will be surprised *most* of the time.

Learn from the Past

Century after century the Romans beat the Gauls. The Romans had the wonderful capacity to eliminate the outmoded, while the poor Gauls never once thought of correcting their errors. They continued to fight the same old way time and again and lost as often. Unfortunately, many corporations are more like Gauls than Romans.

In their analysis of corporate strategy formation in one firm that had changed its strategy a grand total of twice in thirty-seven years, Henry Mintzberg and James A. Waters state, "One conclusion that has appeared in all of our studies is reinforced here. Major shifts in strategic thrust happen only rarely. . . . The fact of only two major shifts in thirty-seven years is not unusual." Other research bears out that strategic change is slow not only in changing to the new but getting rid of the old. If a company has been super at something, its managers will have difficulty divesting that business. It's true in business and true in war: one of the toughest things to learn is when to call it quits.

Eventually *every* strategy proves itself inadequate to meet changes in the company's environment. Conditions, finances, competitor's abilities, prices, technology, and markets are all ever-changing. For this reason, your strategy should never be considered finished or fixed. If you will not win as you are now, either get out of the war or develop what's necessary to win.

"Departures from tradition," writes Rosabeth Moss Kanter in *The Change Masters*, is one of the major building blocks that increase a company's capacity to meet new challenges. And it is possible to depart from tradition and to rise to competitive superiority because of it.

The 840-store Computerland chain is departing from tradition by introducing the concept of "design your own." Consumers can buy a microprocessor just as they do a stereo, component by component. The chain stepped out of the field for this innovation, borrowing the idea from Burger King's "have it your way."

Many individuals and corporations have become strong in an area in which they were originally weak. For example, it's widely recognized that professional public speakers—the Leo Buscaglias of the world—became great orators *because* they worked hard at overcoming severe shyness and fear of public speaking. Craig Bredlove became the holder of the on-land automobile speed record *because* he was originally terrified of driving a car fast.

The British naval strategists eventually became the best in the world *because* the British navy was originally beaten so easily and often. After each loss the causes were scrutinized and appropriate principles, standards, and procedures introduced until the British dominated the sea. Toyota dominates America's imported cars. It didn't always. In August 1957, Toyota brought its first two cars to this country. Popular sellers in Japan, they failed miserably here. Critics called them "overpriced, underpowered, and built like tanks." But unlike the Gauls, Toyota learned from the experience and went back to Japan to come up with a better product. Now Toyota is telling the story in its advertising. According to a Toyota ad:

> We had to start all over.
>
> The message was painful and very clear; if we wanted to sell cars in a place like America, we had to start all over. And make a better car.
>
> So we started over. And worked very hard for many frustrating years.
>
> We stretched our technology farther than we had ever stretched it before.
>
> We tried out ideas that had never been tried before.
>
> We made every mistake that we could possibly make.
>
> And one day we did it: we made a better car.
>
> And the rest, as they say, is history.

In television manufacturing the United States "fell asleep and didn't advance the product fast enough," states Walter K. Joelson, chief economist for General Electric. GE isn't going to let that happen again, not in appliances, anyway. After questioning whether to stay in the appliance business at all and deciding they would, GE has instituted a five-year one-billion-dollar program to upgrade its line of appliances.

While GE learned and changed, one of its competitors lost because it didn't. KitchenAid just could not adjust to the quickly changing dishwasher industry and because of it lost about half its market share in only a few years. In particular, KitchenAid never learned that to compete against GE or Whirlpool, it needed a full product line. Whirlpool, for example, maintained seven models of washers ranging in price from $350 to $750. KitchenAid's run between a narrow range of $500–$750. Also, KitchenAid never filled out its line with products like refrigerators or ovens.

In car wars, American automakers had more than one chance to beat off the invasion of small foreign cars but were blind to the importance of learning from the past. The sequence was this: Volkswagen's Beetle entered the country in 1949, a few years after the revival of European auto manufacturing following World War II. Since Americans bought more cars than all the rest of the world combined, it should have been obvious that the United States would be targeted for European expansion. At the time Detroit held an anti-small-car bias and frankly didn't think the little bugs would pose a serious threat. Supposedly, Henry Ford II said, "Big cars mean big bucks. Small cars mean small bucks."

By 1959 the Beetle had captured 10 percent of the American market. The Big Three fought back, producing their own small cars. By 1962 foreign car sales fell to less than 5 percent. The problem was that the strategy had worked—perhaps too easily. Detroit didn't press its advantage because it still didn't like little cars and thought it had beaten the competition back.

Ten years later the Japanese invaded, and a new batch of so-called import fighters was manufactured—the Vega, Pinto, etc. Again foreign car sales dipped, and again Detroit was satisfied with a victory and again quickly turned its attention to the "real" car business.

By the mid-seventies, when American companies realized that the small-car business was the right car business for the time, they could no longer counter the Japanese companies, which were fully equipped to churn out six million small cars every year, more if needed. Like the Gauls, American companies hadn't learned to correct their errors.

Vary Your Opponents or Methods

Napoleon counseled never to fight too often with one enemy or you will teach him your art of war. For the very same reason, no basketball coach of a fine team wants to play the same team twice in the same season, even if it's a lousy team. Of course, if your competitor is as obtuse as the Gauls, feel free to fight him as often as you wish.

If you must fight the same opponent often, as most companies must, introduce a new twist to your strategy or tactics. In broad outline, for example, Napoleon's methods were always the same, and always involved surprise, one of the principles of war. But he always changed *how* he would spring the surprise. The opposing commander *knew* he had a surprise in store, but he never knew when it would come or what form it would take.

French philosopher Henri Bergson wrote, "A truly living thing never repeats itself." In warfare, one reason the victor stays alive is *because* he avoids repetition. Stay alive.

Never Assume the Next War Will Be Fought
Exactly Like the Last One

It never is; never will be. "The war the generals always get ready for is the previous one," wrote Maj. Henry Tomlinson.

Keep the Plan and Its Execution Simple

*Few orders are best, but they should be followed up
with care.*

Maurice de Saxe

After working hard to develop a solution for a corporate client, I brought the company's managers together in a conference room to hear what I had come up with. I began, "The problem is a complicated one, but I think the way to solve it is very simple. Three things can be done. First . . . , then . . . , then . . . If those three things are done, the problem will no longer exist." In all I had talked for two minutes. I sat down. They didn't buy it.

Two weeks later I went back. Same people, same conference room. This time I presented the same solution but added all kinds of wrinkles, features, and contingencies. It all sounded very complicated. When I finished an hour later, the head of the division slapped the table. "Now we've got something," he declared.

What had dawned on me between the two meetings was that these people needed complexity. Simple solutions—even though perfectly good—didn't have the right ring to them.

Social scientists tell us that human culture generally enlarges and becomes more complicated apparently for no other reason than to be larger and more complex. Beyond a certain point the complexity serves no purpose and actually becomes counterproductive. Isn't this complexity gone out of control, beyond the fundamental requirement of the situation, the cause of considerable corporate confusion and woe?

"There seems to be a widespread assumption," writes John D. C. Roach of Booz-Allen & Hamilton, "especially in U.S. firms, that the more complex a planning process is, the better its results. One unfortunate by-product of this attitude is that the planning process may become an end in itself: highly elaborate organizations are created whose only real product is paperwork."

Business and warfare point to the same conclusion and the same remedy for it. Complicated plans and directions confuse; simple ones can be followed.

Not only do complicated plans confuse your own people—and that is cause enough to demand simplicity—they afford your competition time to counter you. "It is evident," writes Clausewitz, "that a bold, courageous, resolute enemy will not let us have time for wide-reaching, skillful combinations. . . . By this it appears to us that the advantage of simple and direct results over those that are complicated is conclusively shown. . . . Therefore, far from making it our aim to gain upon the enemy by complicated plans, we must rather seek to be beforehand with him by greater simplicity in our designs."

None but the simplest plan succeeds in battle or business warfare.

Look at the plans and proposals before you right now. Even without being able to see them, I'm guessing that 70 percent of them have too many unnecessary contingencies attached to them. Whatever your objective, start by making it succinct enough that you can fit it on the head of a needle. Then start paring it down.

Regroup and Reinvigorate After the First Contact Is Made

"You have a plan. You have an objective. Your men get started with the objective in mind. But in the course of getting to the objective and taking up fire positions, disorganization sets in. The men look for cover and that scatters them. Fire comes against them, and that scatters their thoughts. They no longer think as a group but as in-dividuals. Each man wants to stay where he is."

Earlier in this chapter we quoted Vegetius as having said that the tactical engagement, full of uncertainty, is the moment in which the leader's talents, skills, and experience show themselves to the fullest. This is what he meant. When the objective is threatened, you have but two choices: aggressively rally your subordinates into action or lose the engagement.

Never Renew an Attack Along the Same Line or in the Same Form Once It Has Been Beaten Back

If your assault has been beaten back, you may be tempted to reinforce and try again in the same way and at the same point. Military history teaches that your opponent will expect you to do just that and will himself reinforce that point. If you come back again, trying it exactly as you did the first time, you'll probably lose. Instead, either attack differently or at some other point. In its business warfare, IBM found its American computer retail stores not very profitable. Rather than renewing its attack in that business, IBM simply sold the stores to Nynex Corporation. Be aware that changing the direction—of corporations or individuals—is easier said than done. Once started, an action tends to continue in that direction for a considerable period of time.

The American wine industry will have to choose another line for its advertising attack. Years ago someone predicted that most Americans, like the French, would drink wine with their meals. It hasn't happened. The majority of American consumers rarely, if ever, buy wine, and it's getting worse for the wine industry. The sales of domestic wines fell by 8 percent in 1985. Why? A study done by winegrowers of California suggests that it's because of the advertised image of wine as snooty and affluent and because the task of picking the "right" wine has been made to seem so complicated that it has turned away buyers who don't consider themselves as rich or sophisticated and don't want extra complications in their lives. The industry's job: to take a tack that makes wine appealing to the middle class, the average Mary or Joe.

Expect Friction de Guerre

I am more afraid of our own mistakes than of our enemies' designs.

Pericles

The term "friction," short for *friction de guerre*, has become an integral part of military vocabulary. It means that what you expect to happen often doesn't. Clausewitz describes it as follows:

"Everything in war is very simple, but the simplest thing is difficult. The difficulties accumulate and end by producing a kind of friction that is inconceivable unless one has experienced war. . . . Every war is rich in unique episodes. Each is an unchartered sea, full of reefs."

It is like setting out on a simple business trip. Your car breaks down, or the bridge is down, the road is under repair, and the hotel, full now, hasn't any record of your reservation. So it goes. Friction means all those things that can go wrong will go wrong to make any plans fall short of the mark. Friction is what distinguishes real implementation from implementation on paper.

Empathize with Your Competitor

The highest form of generalship is to balk the enemy's plans. To do it requires the ability to anticipate those plans; and to do that requires an empathy with your adversary. A good general spends a lot of his time putting himself in the other guy's shoes.

Early in World War I the British and French took for granted that on the Western Front the German design was to fight defensively. They were shocked to the core and completely unprepared when, in February 1916, the Germans opened a major offensive at Verdun.

In order never to be caught by surprise, wrote Frederick the Great, "picture skillfully all the measures that the enemy will take to oppose your plans. . . . Then, having foreseen everything in advance, you will already have remedies prepared for any eventuality."

Give the Competitor Some Credit, but Not Undue Credit

> *Experienced military men are familiar with the tendency*
> *that always has to be watched in staff work, to see all*
> *our own difficulties but to credit the enemy with the*
> *ability to do things we should not dream of attempting.*
>
> Sir John Slessor, *Strategy for the West*, 1954

Every good warrior and manager proceeds on the assumption that the enemy's forces are more or less equal to his in ability and spirit. A contempt of the opponent is usually fatal. Napoleon persisted in thinking Wellington a second-rate commander even until the morning of the Battle of Waterloo and despite Napoleon's never having met the Englishman on any field of battle. If ever your managers or staff are thinking, *We're smart, strong good guys, and our competitors are dumb, weak bad guys*, they're falling into the same type of illusion of invulnerability that Napoleon fell victim to.

If you convince your employees that the competition can't stand up to you and yet in the first encounter your competitors resist obstinately, your staff's morale will plummet—along with their confidence in you.

On the other hand, while a healthy respect for the competition is a good thing, undue respect isn't. In all forms of warfare the loser is beaten in spirit before he is beaten in fact. Never let respect become intimidation.

Consider Planning a Process, Not a Thing

Plans should never be solidified, never finalized, for as soon as they are, the opportunity is already gone. Probably the best thing to do would be to pencil in a plan as you do a tentative appointment on your calendar, then order a large supply of erasers.

Don't Abandon the Plan Too Quickly

The Schlieffen Plan was Germany's military blueprint for World War I. It was not a bad plan. As a matter of fact, it was a perfectly fine one. It failed mainly because the German leaders lost the nerve to stick with it. When the Russians advanced into eastern Germany—a move the Schlieffen Plan had anticipated—the Germans rushed reinforcements from the west, thus weakening their forces at the strategically worst moment.

At the first touch with actuality all plans appear wrong or inadequate. If they are, adjust them, but not without first being strong enough to give them their chance. Bear in mind, too, that the first reports coming in should be listened to with a skeptical ear. Usually things are neither as good nor as bad as the people giving you those first reports believe.

Try Once; Step Back

When facing complex, uncertain events, it's best to move forward toward your end objective one step at a time. Implement the first part of the program and work out its problems; then move on to the next. Any good sales representative is well aware of the value of "try once; step back" at the tactical level. If you're selling, and one benefit doesn't convince the prospect, drop it and try another till one clicks. It never matters which benefit you think is best; all that matters is the one his eyes brighten to.

Be Patient with Messiness

Good armies and companies are rarely tidy.

Ask Yourself What Difference It Makes if the Enemy Lays a Few Sieges

In preparing and implementing your plans, realize that not all areas are important ones, and remember that no company can be strong everywhere. Often certain parts of the enterprise can meet defeat without seriously affecting the outcome of the whole because they aren't decisive. But the loss of either too many of them or of a few really decisive points can immediately change the whole situation against you.

Take the Luck Factor into Account

There is no human affair which stands so constantly and so generally in close connection with chance as war.

Clausewitz

Imagine for a moment that you prepare a wonderful project plan. It's perfect. Yet when you launch it, bad luck enters in and screws up everything. You know that the plan was fine, but chance was against it. The odds are that the plan will never be tried again because it has failed. It's said that luck evens out in the long run, but the problem is that in the short run it can run against you.

Luck. Probably in no other occupations do the participants believe in luck as a factor deciding success or failure as in war and gambling. In ancient days it was believed that fortune alone determined who won or lost a war. And even the most scholarly and erudite studies of warfare usually discuss luck. Napoleon considered himself lucky and tried to select lucky subordinates. After listening to a long, detailed description of the capabilities of an officer who was being considered for a position, Napoleon barked out, "Yes, yes, but is he lucky?" Patton code-named his headquarters "Lucky." Lucky Frederick the Great was completely surrounded on all sides by two hundred thousand men and awaiting annihilation when the czar of Russia died, suddenly breaking the coalition against Frederick and immediately establishing him as the winner.

Luck figures in business warfare, too, probably more than we're willing to admit. For fifteen years after it was founded, the publication *Pro Football Weekly* had slowly but gradually captured a hundred thousand-plus circulation. Then, in 1982, the NFL players went on strike, and thirty thousand people canceled their subscription virtually overnight. The publication never recovered, and its management eventually filed for Chapter 11 bankruptcy. Take an uncomplicated transaction like a sales call. Which representative will call first? Will the prospect be in? Will the mail be delivered in time? Will the secretary give the message or be out to lunch? Will the prospect be in a good or bad mood? Even such a simple situation has a great range of possibilities or chance outcomes. Simply tracing the performance of the stock market on days of the week reveals that the worst day of the week far and away is Monday. Why? The time afforded by the preceding two-day weekend for the chance event to occur.

Experiences of famous commanders tell us that luck is felt to be "in" the person, not in the events, that some people are luckier than others and that great leaders are typically lucky; that in short operations such as battles or brief corporate projects it is essential to take the factor of luck into account; and that your plans should be based on the expectation that LUCK WILL RUN AGAINST YOU.

CHAPTER FOUR

Concentrating Greater Strength at the Decisive Point

The principles of war could, for brevity, be condensed into a single word—"Concentration."

B. H. Liddell Hart

In March 1796 Napoleon was appointed commander in chief of the French army in Italy. He left Paris after a two-day honeymoon with Josephine and joined his army on the Italian coast. He found his forces in a condition in which managers often find the subordinates left to them from an old regime—disorganized, scattered, uncoordinated, low in spirit, dissatisfied with their pay, located in a terrible strategic position, and fronted by an enemy almost twice as strong and getting set to attack.

Like any savvy leader who is a stranger to his new army, Napoleon quickly established a vision of the greatness that was possible, if the soldiers followed him. He promptly proclaimed, "I will lead you into the most fertile plains on earth. You will conquer rich provinces and large towns; there you will find honor, glory, and wealth." Then, understanding that leaders who are followed really do put their money where their mouth is, he borrowed the funds to pay his men.

Two armies faced Napoleon's army of thirty-seven thousand—the Austrians with thirty-five thousand men and the Sardinians with twenty-five thousand. Extraordinary leaders often do the unexpected, and Napoleon was extraordinary. Before his career would end, he would fight more battles than Alexander, Hannibal, and Caesar *combined*, and history would call him the greatest strategist of the Western world. That day at Montenotte he had every reason for staying on the defensive, but being extraordinary, he ordered an immediate attack. His first executive order was for his officers to do what his name is forever associated with—to CONCENTRATE their troops, to mass them at one critical point of the action. His plan was daring: to concentrate his army at the point where the two enemy armies joined and to attack quickly, drive a wedge between them, and turn each back to its camp.

He ordered the French forward in four columns, one to move against the Sardinians and three to concentrate on the space between the armies and to attack one small unit of the Austrian army that had been left to defend the area. The advance succeeded. He split the two enemy armies and beat the Austrian defenders in the area.

Napoleon was never one to dawdle. Long wars and long engagements are expensive; short ones are cheaper. He always tried to reach the objective as quickly as possible and preferred quick, short, snappy wins. So, the morning after splitting the enemy's armies, he swung his army first against a Sardinian detachment and later in the day against the Austrians. When Napoleon's troops came hurtling forward, the Sardinians threw down their arms and surrendered. In the afternoon he turned his army around and sent it against the Austrians. That started three days of intense fighting. Throughout, Napoleon applied his style of warfare—concentrating more force than the enemy at one point; winning there; then shifting to another point, piling on the force; winning there; reconcentrating, etc.—and always adding to the force of his troops by moving with lightning speed and pounding the enemy with artillery.

Soon the Battle of Montenotte was history and Napoleon had won his first campaign—an impressive and daring beginning for a short, ordinary-looking young man of twenty-seven with absolutely no previous experience in commanding an army.

In a statement to a friend about Napoleon's art of war, French general Ferdinand Foch (1851–1929) declared:

> . . . it seems to me that his art consisted in a few principles of extraordinary simplicity and clarity. These he used with the touch of a master. To husband his troops; to use them judiciously so that the enemy might be attacked at his weakest point with superior forces; to keep control of his men, even when they were scattered, much as a coachman holds the reins, so that they could be concentrated at a moment's notice; to mark down that portion of the opposing army which he aimed at destroying; to discern the critical point where defeat might be turned into rout; to surprise the enemy by the rapidity of his conceptions and operations. . . .

On a spring day in 1921, a century and a quarter almost to the day after Napoleon's victory at Montenotte, Alfred P. Sloan, Jr., of General Motors sat down with a group of other GM executives. Their objective? To design a strategy for attacking Ford Motor Company's dominance of the low-priced-car market. As of this writing, GM is the world's largest corporation, with sales of $24.6 billion. In 1921 GM wasn't even first in the auto industry. Ford was number one, producing half of all American cars. GM was number two, but far behind, producing only 10 percent of American cars.

At the time, Ford made just one car, the low-priced Model T. GM produced a number of lines, from the medium-priced Chevrolet to the high-priced Cadillac. In a sense, GM and Ford weren't even in competition with each other. *Both* Henry Ford, Sr., and Sloan realized the same thing—the future of the industry was in the low-priced market. Ford was deeply entrenched there; GM didn't have a foothold. Like the French under Napoleon, GM decided to waste no time in attacking. The dominance of the Model T was, to use Foch's words, "the portion of the opposing army" that GM aimed at destroying. And the company would do it by concentrating superior force at that critical point—the low end of the price range.

At the time, Sloan wasn't the head of GM. Pierre du Pont was president and CEO, but it was Sloan's concentration strategy that GM adopted. The war was fought over six years, but the major strategic moves can be described quickly.

Sloan divided the entire auto market into six price classes, from $450 to $600, $600 to $900, etc., up to the sixth class, $2,500 to $3,500. Chevrolet was designated as the division that would advance directly on Ford's Model T low-priced market. Chevrolet concentrated only on that market; it didn't make a car for any other price range. Dividing the market into price classes was the first of Sloan's strategies. Like Napoleon's, it was based on concentration of forces. Napoleon's main concentration was the point between the enemy's armies; Sloan's was Ford's main "army"—Ford itself, the Model T.

When GM attacked with its low-priced Chevy, Ford countered by lowering the price of its car. Initially, Chevrolet lost money, selling only seventy thousand cars; Ford sold more than a million. The next year, 1922, Chevrolet began to gain a foothold with consumers and sold two hundred thousand cars; Ford did better, selling over one and a half million.

Chevrolet introduced innovations that differentiated it from the Model T. Henry Ford considered the Model T invincible and modified it only slightly. In 1923, with expanded dealerships, Chevrolet sold 450,000 cars. From that point on Ford had no hope of driving GM from the market.

In 1924, while Ford was regaining ground and setting record sales, Sloan—now GM's CEO—concentrated his effort on upgrading the product. In war, an innovation—if it's the right innovation at the right time—changes everything. A major innovation creates new rules. GM's new "K Model" changed the rules of the low-priced-car war. The "K Model," introduced in 1925, could be either an open or closed car; the Model T was essentially an open car. GM's decision to concentrate on product upgrading couldn't have been timed better. In 1925, 56 percent of all cars sold were closed; 73 percent were closed cars in 1926; 82 percent, in 1927. GM now brought the price of its closed K cars down and down until it was approximately the same as Ford's. Unable to compete on the basis of price or product quality, Ford was forced to completely shut down its factories for almost one entire year while it designed the Model A to replace the T. With Ford out of the market in 1927 and 1928, Chevy's sales skyrocketed to over one million cars each year.

Ford rallied briefly in 1929, its fine, four-cylinder Model A out-selling Chevy. But again Sloan counterattacked through concentrating on innovation. Chevy engineers quickly developed a low-priced six-cylinder design that brought GM victory.

The president of a small training company told me recently, "We've decided to offer only sales seminars. It's what we do best, and hell, we can't make a dime doing thirty-five different kinds of programs." That's an example of concentration, no different in essence than Napoleon's or Sloan's.

CONCENTRATION. In field after field, the one who concentrates wins. Athletes know its value. "Concentrate," says former football great O. J. Simpson. "You've got to concentrate." The rich know it, too. The number-one characteristic of materially successful people is not a special talent that they possess that others don't but their ability to focus totally on an enterprise and to pour their energy into it day after day after day.

Artists learn the value of concentrating, too. Marvin Hamlisch is the winner of three Academy Awards (in one year) and composer of *A Chorus Line*, the most successful musical in Broadway history. He told me that when he was a child he had to choose between concentrating on music or playing baseball and other games after school. When he was older, he had to decide between concentrating on composing or being a concert pianist. He chose music, then narrowed his concentration to composing.

Napoleon and Sloan faced major concentration decisions, too. They could have thrown their resources *anywhere*. They won because they selected the best place, the decisive point.

All great military leaders down through the corridors of time have come to the same conclusion—that a victory comes more easily if strategy and tactics are based on some form of concentration at a decisive point. Tamerlane (1336–1405) said, "It's better to be on hand [at the decisive point] with ten men than to be absent with ten thousand." Clausewitz added, "The theory of war tries to discover how we may gain a preponderance of physical forces and material advantages at the decisive point." And Ralph Waldo Emerson called concentration "the secret of strength in politics, in war, in trade, in short in all management of human affairs." In business warfare, concentration is the implementation of someone's or some groups' decision TO BRING TOGETHER GREATER STRENGTH THAN THE COMPETITOR AT A SINGLE DECISIVE POINT WHOSE POSSESSION PROMISES COMPETITIVE ADVANTAGES. "Greater strength" means having more resources or better ones than the competition's at that point. "Decisive point" refers to the critical area or issue, the one that can exercise a marked impact on the result of a war, a campaign, battle, or project.

The Side That Concentrates Better Wins

In the movie *Seven Samurai* there is a wonderful scene in which a great swordsman refuses to duel another man because he knows, without having to fight, that he can beat him. The actual fight would only validate what the warrior already knows. As far as he's concerned, the combat isn't necessary. The man insists on fighting, however, and as expected, the great swordsman cuts him down.

Identifying the decisive point on which you'll concentrate your manpower, material, time, money, brainpower, advertising, production, sales, etc., is very much like that movie scene in that accurately choosing that "point of points" is tantamount to having won already. If you succeed in choosing the strategically or tactically decisive point in your enterprise and, having found it, follow the principles of competitive warfare, the actual competition should only confirm the wisdom of your choice. You should be proved right.

The competition within the competition of business concentration is the contest for strength at the decisive points. Specifically, if we cast an eye at business warfare, we see exactly what we see when we look at military wars. Winners concentrate; losers don't, or not as well. Managers in competitively superior companies:

1. Select the decisive point in the competition.

2. Concentrate superior forces there.

The losers—losing managers, losing companies:

1. Either don't concentrate forces or

2. Concentrate them at the wrong points or

3. Concentrate at the right points but with inferior forces.

For instance, in the GM-Ford example, both corporations selected the right market—low-priced cars. In fact, Ford chose it first and was well entrenched when GM entered the battle. Given that market, however, price alone was not all that was decisive. GM concentrated its resources on what turned out to be the decisive point—innovation. Ford didn't, and GM won.

Using Force at the Decisive Points

Some military masters have called the maintenance of the objective the most important principle of war, and Clausewitz said that next to victory the subprinciple of pursuit (see chapter 5) mattered most. But if a poll were to be taken of all the great military masters of history as to which principle was most important, there is little doubt that the majority would cast their vote for the principle of concentration. It has been referred to as the GREAT principle underlying all operations of war and the one that MUST be followed if the combatant is to succeed.

It's easy to bandy the word "concentration" about. All a manager has to say to subordinates asking for advice on most anything having to do with increasing productivity or achieving marketing superiority is one word, "concentrate," and that manager will eventually be proven a "genius."

Strategically and tactically, to "concentrate" means:

Strategically:
1. Through strategic movement, throw the bulk of your forces successively upon the decisive points of a theater of war.

2. Maneuver so you engage fractions of the enemy's forces with the bulk of yours at that point.

Tactically:
3. Maneuver so that you throw the coordinated mass of your forces upon the decisive point.

4. Make certain that you not only commit your concentrated forces to the decisive point but that you do so quickly and at the most opportune time.

There is no reason to believe that Alfred Sloan of GM was aware of the military principle of concentration, but certainly his strategy of dividing a market into price classes fits the four steps of concentration listed above. Skillful segmentation—dividing a broad market (a glob) into more distinct, finer, narrowly focused markets—is a concentration strategy that worked for GM, is still working today, and, if trends continue, will still work many years from now. Napoleon won by driving a wedge between armies; corporations often win by wedging into market segments more skillfully than their competitors do.

For years it was all quiet in the seven hundred-million-dollar-a-year baby-food industry. Little was spent on advertising, and innovations were considered unnecessary. After all, baby food is just . . . baby food. Suddenly, in 1984, Beech-Nut took the offensive against Gerber, the industry leader. Beech-Nut segmented the market into four distinct ages of babies and introduced four different groups of products. According to Beech-Nut, the segmentation strategy has cost Gerber a 12 percent market share in two years.

It may seem surprising that Ford offered just one type of car in the 1920s. You might be thinking that's old-fashioned, but as late as 1984, BIC sold just one type of disposable lighter. Scripto attacked, wedging in, using the very same strategy that GM used more than half a century ago. Scripto has created a number of throwaway lighters, some models more expensive, some less costly; some for women, some for men, etc.

Concentrating Strength Against Weakness

You're the general. You look out over no-man's-land, and across from you you see the enemy's line. The line is of unequal thickness. There are stronger, heavily protected points, and there are weaker, thinly guarded points, too. Which do you attack?

You may decide on a concentrated, fully coordinated blow aimed at a stronger point. You may even go after his strongest point of all, arguing that if you break him there—at what Clausewitz would call his "center of gravity"—he will fall and that all you'll have to contend with later on is meager resistance from the weaker elements of his army.

On the other hand, you might be thinking you could lose by attacking where he is strongest. Or you might figure that an attack at the enemy's strongest point will be expensive and exhausting regardless. You may not win, but even if you do, you may be so weakened that the remaining units of the enemy's army may be strong enough to annihilate you. Or a third army may wait on the horizon until you and the enemy exhaust each other and then swoop down and destroy you both.

Or you may choose the weakest point in the enemy's line and attack there. Erich Ludendorff (1865–1937) was the German general who invented battlefield tactics as we know them today. He recommended concentrating the attack where it is easiest to penetrate. The trouble with this is probably apparent to you. That point in the enemy's line may be weak simply because it is tactically or strategically unimportant to him. By occupying it, you may gain little, if anything, of value. Your competitor may even be willing to give it to you without a fight. Neither Sears nor any other reasonable retailer locates a store simply because a competitor isn't located there. The "easiest" place, but certainly not the best, for retailer penetration would be an empty prairie.

Almost always your best bet is to concentrate your strength against the opponent's weakness, but at not just any point, only the one that's in fact decisive. This corresponds with a winning business-warfare-concentration formula that probably accounts for as much as two-thirds of a company's success: COMPETE IN WAYS THAT MAXIMIZE YOUR STRENGTHS AGAINST THE COMPETITOR'S WEAKNESSES IN THE MOST OPPORTUNE MARKETS AND WITH THE RIGHT PRODUCTS. Estimates based on data from hundreds of companies confirm that firms that follow this type of concentration will do approximately twice as well as companies comparable in size and type of markets that don't follow it.

Your concentration must be pitted against the competition's dispersion. Mass where the other side is strung out; invest heavily where they invest little; place a high priority on what they don't emphasize.

Never concentrate your forces at a point that isn't valuable to you even if the opponent is weak there. Going for the thinnest place in the hedge makes good sense only if what's on the other side of the hedge appeals to you.

If the competitor is only relatively strong at a point that's of crucial importance to you and you can afford it, mass greater forces there, then attack.

If you and he are of approximately equal strength at the decisive point, neither of you is concentrating your forces.

If he is more highly concentrated than you at the point that's crucial to you both, usually you'll have to go on the defensive, at least temporarily, or swing around and concentrate at another point.

THE CONSUMER IS THE ULTIMATE
DECISIVE POINT OF CORPORATE CONCENTRATION

In his well-known book *The Structure of Scientific Revolutions*, Thomas S. Kuhn describes how each age creates paradigms, or conventional models of reality. These are often so widely agreed on that we accept them as reality, forgetting that they are only models, only representations of reality that can be changed. "Paradigm shifts" occur when the model we hold is displaced by another. Kuhn indicates how major paradigm shifts occur in scientific fields, but they need not be so colossal. A paradigm shift that my parents have not adjusted to yet is that poetry doesn't have to rhyme anymore. Parents often can't stomach the music their teenagers enjoy because of the shift in the definition of what music is "supposed" to sound like.

A corporate paradigm shift of monumental proportions occurred in the 1950s when marketing replaced selling as the principal orientation of most businesses. Corporations stopped concentrating so much on using promotions to convince, persuade, or cajole the consumer into buying the product the company had already manufactured. Instead, they tried first to understand consumer needs and to manufacture a product that could satisfy them, then to emphasize selling.

It was 1960 when Theodore Levitt's article "Marketing Myopia" appeared in the *Harvard Business Review*. That piece was extremely influential in that it both described and accelerated the trend toward marketing. Levitt described the essential difference in the old-style sales orientation and the new-style marketing orientation as follows: "Selling focuses on the needs of the seller; marketing on the needs of the buyer. Selling is preoccupied with the seller's need to convert his product into cash; marketing with the idea of satisfying the needs of the customer. . . ."

This is now old hat. But in its day the initial paradigm shift toward consumer satisfaction above all else was revolutionary. It meant that peddling was out and closeness to the customer was in. It also meant that in military terminology TERRAIN was the new decisive point of corporate concentration. Gaining the consumer came to be the equivalent of wars fought to take and hold a hill, frontier, capital, or section of a country.

That was the style of fighting military wars in the eighteenth century, up to 1780. The object of most warring in the 1700s was to acquire terrain, not necessarily to defeat the enemy in battle. Every commander must decide if the objective is to go after (1) terrain or (2) the enemy's army. Most eighteenth-century generals chose the former. The general who said, "I see only one thing, the enemy's main army," was *not* speaking as an eighteenth-century general. While some companies set out to knock off the other company's army, most don't consider that the main priority.

For example, in spring 1985 Xerox came out with a number of new computers, printers, and software programs. Rather than "seeing only IBM's main army" and going toe-to-toe with IBM in retail stores, Xerox concentrated on a different terrain—large offices requiring visually oriented corporate publishing equipment. This is good, sensible twentieth-century marketing.

The overwhelming majority of corporations compete *with* other companies, but *for* the consumer, the consumer being the main objective. Like the geniuses of eighteenth-century warfare, twentieth-century corporate executives and managers regard the objective of the collision of competitors to be largely a matter of winning territory. Terrain-seeking generals apply the principle of concentration to gain and hold specific plots of ground, territory, land—terrain in the literal sense. Corporations apply the principle to gain their own "terrain"—consumers who will purchase what the company wishes to sell.

CORPORATE TERRAIN AND DECISIVE POINTS

The nature of the ground is the fundamental factor in
aiding the army to set up its victory.
Mei Yao-Ch'en (1002–1060)

During the past quarter of a century plus, corporations have been following these four axioms of profitable marketing concentration:

1. Segment the market.

2. Find a niche (decisive point of concentration) you can operate in successfully.

3. Learn what consumers in that niche want and give it to them at a profit.

4. Differentiate yourself from other firms.

Each of these axioms is based on the immense importance of taking and holding consumer terrain.

The marketing paradigm shift is far from over. It is still under way. It is evolving in the direction of increasingly narrow forms of corporate concentration on more and more finite consumer decisive points. Analysts of emerging trends are taking note of (1) the increasing diversity in life-styles of consumers; (2) a decline in traditional mass-marketing strategies because of inefficiencies in adequately addressing that diversity; and (3) the proliferation of strategies of extremely narrow microniching. The trend toward concentrated microniching is so pronounced as to border on being revolutionary. In military terms, particularly competitive corporations are pursuing the advantages of massed assault at very narrow decisive points. Antoine Henri de Jomini said the ability to select decisive points in the combat is "the greatest talent of a general, and the surest hope of success." Napoleon was just one great general who possessed this talent to an amazing degree, reportedly being able to select crucial points of a battlefield "at a glance."

Below are examples of decisive points some of the most successful corporate combatants in American business warfare are selecting in their attempt to win the competition for superior concentration. They demonstrate that, like Napoleon, managers, too, can possess the greatest talent of a general.

Age as Decisive Points

Americans over fifty years of age make up approximately 25 percent of the population but hold a whopping 50 percent of the country's disposable income. Since this is the fastest-growing age segment, it should not be surprising that growing numbers of companies, including many of the very largest, are seeing late-middle-age people and senior citizens as critical points of concentration.

Sears, the country's largest retailer, has developed its Mature Outlook Club specifically to attract customers in the fifty-plus segment with discounts of up to 25 percent on everything from eyeglasses to garden tools after an initial membership fee of $7.50.

Nursing-home care is probably the largest unmet need that insurance was designed to help. Estimates are that less than one hundred thousand people carried long-term care insurance in mid-1985. Yet the potential for such policies within the next five years alone is projected to be three to five billion dollars.

Pearle Health Services of Dallas—Pearle Vision—is the nation's top seller of prescription eyeglasses and contact lenses. One age segment Pearle is particularly interested in is made up of the baby boomers nearing their forties. First of all, that's a huge segment of the population, and secondly, it's in the forties that many people start losing their 20/20 vision. Teeth and gums go bad at that age, too, so Pearle, a marvelous merchandiser, is moving into retail dentistry also, operating, initially at least, out of Sears stores.

At stake in the competition between Mars and Hershey is which will win the battle of the sweet tooth in the country's eight-billion-dollar-a-year candy market. Between them these two giants produce all ten of America's best-selling candies—five each—and share a huge 70 percent of the candy-bar market. The next largest competitor, Peter Paul Cadbury, sells in the vicinity of 9 percent of U.S. candy bars.

Although only one-third of the total size of Mars (which has businesses other than candy), Hershey, more innovative than Mars, is making Mars uncomfortable with its age-based concentration strategy. It's developing candy products that appeal to adults, the 55 percent of candy eaters over eighteen. Since 1981 it has introduced two new bars that are less sweet and "chocolaty" than kiddie bars. It has also introduced a Big Block line of fifty-cent bars that, being thicker and offering more to chew, appeals to adult—particularly male adult—mouths.

Winnebago is the second-largest manufacturer in the highly competitive mobile-home industry, which has seen the demise of one-third of the competition in the last five years alone. Winnebago has succeeded in cutting out a very narrowly concentrated niche that it alone occupies and will have all to itself for some time. It began with Winnebago's decision not to pursue any of the far-flung ventures it had been considering but just to stick with the company's core strength, making mobile homes. Next came a heavy investment in research to locate the most promising niche and a push to increase productivity and improve product quality. Winnebago selected its terrain on the basis of age. Two-thirds of mobile-home buyers are forty-five or older. Winnebago didn't see this as its market. Instead, it targeted on younger buyers, the thirty-five- to forty-four-year-olds whose numbers, research shows, are expected to climb by 40 percent in the late 1980s.

Regional and Geographic Points

A product may flourish here and fizzle there because of differences in regional decisive points. The trend toward regional marketing has been credited to the book *The Nine Nations of America*, in which author Joel Garreau points out the existence of nine distinct "nations" within North America, each nation holding a population sharing distinct values, attitudes, and styles.

Why do people in Seattle buy more toothbrushes per capita than residents of other U.S. cities? Why is Salt Lake City the number-one market for candy bars and marshmallows; or Dallas/Fort Worth number one in popcorn sales; and Philadelphia tops in iced teas? Who knows? But what is known is that more and more companies are becoming advocates of regional marketing. For example, S. C. Johnson & Son, makers of Raid, closely studied the regional buying habits of consumers. In just one year of concentrating on regional marketing, it increased its market's share in almost all of its regions (sixteen out of eighteen) and raised its total market share of the $450-million-a-year U.S. insecticide business by 5 percent. Raid is raiding—regionally; and its corporate managers are taking quite seriously Machiavelli's advice: "A general should possess a perfect knowledge of localities where he is carrying on a war."

Scripps Howard, the newspaper and media chain, is attempting to rejuvenate its corporate vitality by switching geographic points. It is putting money into its urban and suburban dailies and is lessening its concentration in smaller markets by selling many of its nondailies.

Brokerage house Edward D. Jones & Company has more offices than any other stockbroker in the country, 970, many of them just one- and two-person operations. The company's concentration strategy is small-town America, generally neglected by other brokers. The others focus on New York as a decisive point, or Los Angeles. Jones & Company concentrates on the Spearfish, South Dakotas of the nation. Yet with more than three thousand people working for it, Jones & Company has more employees than First Boston Corporation.

In summer 1985, California Cooler launched its Midwestern advertising campaign for its wine-and-fruit drink. In the ad, a bar patron gripes about Californians, mentioning their flakiness, but orders a California Cooler, anyway, because it's good. Although done tongue in cheek, the ad masterfully blunted the objection of those in the nation's central states about buying anything Californians make or like. In the minds of Midwesterners, who think of themselves as practical and down-to-earth, Californians are hot tubs and tans and strange.

Many competitively successful corporations are concentrating their advertising dollars on regional rather than national decisive points, more by the day. Local events, local personalities, local color, and local tastes advertised on local television are part of the noticeable swing to the regionalization of advertising. According to the Television Bureau of Advertising, the national networks' share of advertising has fallen 5 percent since 1980.

In the past, demand for hotel rooms had exceeded supply. But since 1978 occupancy rates have fallen approximately 7 percent. Having saturated the traditional downtown and inner-suburban markets, hotel chains such as Hyatt and Marriott are looking elsewhere for growth. They are following the population to smaller cities and the outskirts of metropolitan areas.

"The opportunities on the blade side really lie in new geography," says Rodney S. Mills, an executive vice-president for Gillette's international arm. Sensing an opportunity in the fact that only 8 percent of Mexican men who shaved used shaving cream, in 1985 Gillette test marketed an inexpensive shaving cream in Guadalajara. Today, 13 percent of the men in that city use shaving cream, and Gillette is planning to sell its new, specially designed product Prestobarba, "quick shave," throughout Mexico, Colombia, and Brazil. In addition, it is tailoring a marketing program to educate third-world men to the advantages of a smooth face. It's sending out its representatives to demonstrate how shaving is done—to the extent of lathering up villagers while others watch.

Life-Style as the Decisive Point

In specialty magazines publishing today, "affluence" is increasingly the narrow critical point of concentration. More Americans are better off financially than ever before. American Airlines' *Private Clubs* magazine is expected to attract readers with an average household income of $144,000. Even mass-circulation magazines like *TV Guide* are noticing an increased affluence in their readers.

Another life-style decisive point is the busy single adult. "Rent-a-Wife" services have sprung up, the "wife" sitting in for the single, usually a woman, by waiting for deliveries, repair people, doing some of the shopping, etc.

Atlanta-based Aaron Rents, Inc., is nearing one hundred million dollars a year in revenues, almost half its income coming from renting home and office furniture. Futurists paint a picture of a national population on the move, and this trend toward mobility makes furniture rental an ideal business for the late eighties and nineties. Another factor is divorce. When and where the divorce rate is high, the terrain is right for furniture rental.

Douglas Schumann found a terrain niche no one else was filling nine years ago when, during a trip to New Orleans, he overheard someone say it was a shame that New York restaurants didn't have New Orleans crayfish. He asked New York chefs, and they said they would buy crayfish if available. Schumann found a supplier in Louisiana, and in a few months was sending three thousand dollars' worth a week to New York. At present O. D. Schumann has evolved into a major importer of fresh Dover sole, European-grown mushrooms and vegetables, Hawaiian prawns, and other delicacies. In 1985 the company's sales reached $3.5 million, with a 30 percent annual growth expected.

International Decisive Points

Corporate strategic consultant Kenneth Ohmae, managing director of McKinsey & Company's Tokyo office, sees winning multinational corporations pursuing more narrow international concentrations. Says Ohmae, "The new global enterprise will be more deeply and strategically involved in fewer countries, choosing a few and getting to know their institutions and leaders well."

Retailing Decisive Points

"Microspecialization" will be the retailer's concentration buzzword of the late 1980s and the 1990s. It will be based on clearly differentiating between consumers and then a precise targeting of merchandise selection, advertising, and even store design to appeal just to that terrain, that group. To win the retail war, store operators will have only two options in the next few years: achieve superiority in a specialty niche or become dominant in a geographic zone.

"Gadget stores," emporiums selling neon telephones and other wacky items, win by differentiating themselves from competitors by concentrating on the decisive point of the consumer's sense of humor. Recognizing the movement toward specialty buying, in 1982, Sears, the pioneer of catalog merchandising, began marketing its "specialogs," catalogs targeted at small market segments.

Perhaps the ultimate in retail high microspecialization and concentration along a narrow front is the "category killers," stores carrying huge quantities of merchandise of one particular type—toys, for example, or records or furniture.

Most of my family's athletic equipment is bought at a sporting-goods category killer, Sportmart in Chicago. Five times as spacious as its typical competitor, Sportmart stocks seventy types of sleeping bags and fifteen thousand fishing lures, for example. Category killers are named after their marketing strategy: defeat the competition by out-concentrating him. Buy in extraordinary quantities, sell cheap, and by satisfying all the consumers' needs for a particular category of merchandise, keep the customer in your store and out of the competitor's. Napoleon could not aim for a more total massing designed to overwhelm the opponent.

FREDERICK THE GREAT'S
MARKETING ADVICE TO SMALL COMPANIES

"A general should choose his ground with regard to the numbers and types of his troops and the strength of the enemy," writes Frederick the Great. "If he is stronger . . . he will seek the plains . . . because his superiority gives him the means to envelop an enemy on open ground. . . . If, on the contrary, you are inferior in numbers, do not despair of winning, but do not expect any other success than that gained by your skill." It is necessary, Frederick adds, for the smaller army to seek country on which the enemy "would not be able to face you with a front superior to your own."

Frederick is right in pointing out that if you're the smaller army, you should seek a different kind of country than your larger foe. The larger army seeks the "open plains"—national product, national distribution, national advertising, for example. The smaller army is unable to win whenever the larger army faces it with a superior front. It tries to find areas where the larger army's front is not superior. In general, large corporations have the advantage wherever there are economies of scale involved. The companies with the larger "armies" have a virtually insatiable appetite to dominate market share. They *must* increase production so as to drive down the unit cost. If you're a small company, your best path to profitability is to concentrate on business where economies of scale don't matter that much—a highly specialized niche, for example. When Frederick said that the smaller army shouldn't expect any success but that gained by its skill, the skill he meant was the skill of identifying the right kind of country. In business warfare this translates into the skill of the smaller force of identifying a specific promising market segment that is too specialized (too narrow, too esoteric, too different, etc.) for competitors to compete in, especially market leaders.

Business Week, Fortune, and *Forbes* have the big-business terrain. *Inc.* magazine chose not to face them there but to seek country where they were not superior—the small-company market generally ignored by other business magazines. Frederick the Great would have been pleased. And founder Bernie Goldhirsh was when *Inc.* impressively turned a profit in only its second year.

Goldhirsh's winning strategy brings up an issue you should be aware of. During World War I the Allies sent their strongest armies to attack the front where the Germans were strongest, while the weakest German fronts were left to the most poorly equipped armies. The result was a stalemate everywhere. The lesson is that unless you are truly superior to the competition, concentrate your strength against its weak front and leave only a weak force or no force fronting its strength. It's exactly what *Inc.* did. That's not only what the military masters would tell you but the advice of most venture capitalists. Invest in the business with the soft competition.

TWO STYLES OF FIGHTING

In a crucial situation a pitcher in baseball thinks one of two ways, either "I'll throw this guy my best pitch" or "I'll throw the pitch I think this guy can't hit."

Generals of armies and corporate managers think the same way. Some choose to go with their best pitch. They are fighters from a core strength (FCSs). Others concentrate more on the opponent's weakness. These are AVs—attackers of vulnerability. The goal is always the same—to concentrate at the decisive point of some niche and gain a foothold in it. But some companies choose to reach the goal by moving along lines of their own strength. Others elect to slice into the niche by exploiting the competition's vulnerabilities. FCSs and AVs are traveling up the same hill but taking different paths.

Fighters from Core Strength (FCSs)

After an NFL game a TV interviewer commented to a huge defensive lineman that he had seemed to be able to blow away the offensive player trying to stop him. The lineman replied, "Right, strength isn't my weakness." It isn't IBM's weakness, either. IBM has weaknesses, just as any company does, but its core strength *is* its strength, its power, its enormous resources. It fought from its power when, like a great army that could exercise its will with impunity, it took on Apple, the then leader of the personal computer business, and won. IBM didn't even bother with maneuvers at all but went toe-to-toe with Apple, duplicating exactly what Apple had done to achieve its success —its network of retail outlets, its dealer arrangements, training program, etc.

The question the fighter from core strength asks continually is not "What kind of war can the enemy fight?" but rather "What kind of war are we best at fighting?" As a matter of fact, if the core strength is significant enough, the company may not concern itself with what the competition does. Secretaries, Inc., is an office-personnel placement company that's the best in its market. Its vice-president, Joel I. Cohen, told me, "We're number one. Frankly, we don't care what the others do." Core-strength fighters play their own game, not the competition's game.

State Farm is America's largest mutual insurance company and our number-one automobile insurer, with twenty-five million auto policies. While its competitors jockey for positions as full financial service providers with brokerage services, employee-benefit consulting, and money-market funds, State Farm stands clear, selling insurance to the mass public. Nothing fancy, no extras, just insurance.

"We are a very focused company as to what we're all about," says Edward B. Rust, Jr., State Farm's president and CEO. The company's "concentrate on insurance" approach carries on a tradition. Rust's father, Edward B. Rust, Sr., who headed the company until his death in 1985, said that since only a small part of the market was interested in those other services, State Farm would be content to work on the 95 percent while other insurers scramble for the 5 percent.

Whether warrior or manager, fighters from core strength focus on a capacity they possess that the opponent doesn't—at least not to the same degree. The core strength of Wellington's army was Wellington himself. A master of tactics, he chose to fight tactical wars. Roman strength lay in organization—their packed phalanx. The core strength of the Mongols, the finest cavalry the world would ever see, lay in what we would call their human resources, particularly the sheer toughness of the Mongol horseman, his ability to endure terrible hardship and to keep moving no matter what. Strategically and tactically, FCSs focus on fully exploiting that special capability or cluster of capabilities that distinguishes them from the other side and that leads to success on the business battlefield. They strive to be strong in an ability in which the opponent is weak. They are not unique for the sake of uniqueness but because uniqueness in that decisive area wins!

In every industry there are certain tasks or functions or small clusters of them that must be performed extremely well if a corporation is to achieve competitive superiority. Firms that achieve superiority in these areas can substantially increase their chances of winning. Most companies emphasize excelling in product development or marketing, but most any profit-related area can be or become your core strength, including financing, service, and human-resources capabilities. In the oil business the winner simply used to dig more holes than losers. Currently exploration is too risky. Exxon, the world's second-largest company, finds that the best strategy is buying up competitors' producing properties as the market falls. In the liquor business, the vital core function is distribution.

A core-strength concentration often starts with identifying what makes for success in the business. If success in your industry or trade is highly dependent on service, service away. Bask the consumer in it. If the capability that brings success in the industry is sales, focus on selling, etc.

The decisive point that differentiates you from the opponent may follow an innovation in technology. Gustavus Adolphus created added strength in his army by making artillery mobile and coupling it with infantry and cavalry. While the enemy's pieces were fixed into place, unable to move as the scene of battle shifted, Gustavus Adolphus was able to wheel his about freely, a tremendous advantage. Innovating your way to a core-strength superiority can help you leapfrog over competitors.

The core strength may involve a breakthrough innovation but need not. You can always acquire it. In *The Art of War*, Sun Tzu wrote that in war the best thing of all is to take a country whole and intact. The easiest way, if you have the money, is simply to buy a core strength you lack.

The Walgreen Company, headquartered in Deerfield, Illinois, has more than a thousand drugstores nationally but only a few in New York. Insiders in the drugstore business believe that Walgreen's is so strong financially that no market is too large for it. After three years of patiently testing the Big Apple terrain, Walgreen's is leaning in the direction of buying out a competitor with a presence there, then turning on the Walgreen's success formula of intense advertising and high sales volume.

IBM hadn't acquired a company in two decades before buying Rolm, a leading producer of communications equipment. After trying unsuccessfully to become a force in the communications-equipment field for more than ten years, IBM solved the problem by buying a concentration in that business, thereby positioning itself to battle AT&T for the major share of the information-processing market.

A corporation may have to divest itself of other businesses in order to return to its core strength. At the present time, increasingly more firms are pruning off unrelated businesses after unsuccessfully attempting to integrate and manage them, often finding that what Machiavelli wrote is still apropos: nothing is more difficult or dangerous than to initiate a new order of things. The company that chooses to acquire unrelated business will probably at some time or another have to stop to tend to the painful process of mix fixing.

Under the leadership of Martin S. Davis, Gulf & Western cut itself in half, eliminating a crazy quilt of low-profit businesses and forging itself into a leaner company with assets of $3.7 billion in the entertainment and communication business and $7 billion in financial services. In a report, brokerage Shearson Lehman Brothers, Inc., said,

"Once aimless, this diversified company with many capital-intensive businesses has been refocused (under Mr. Davis) into a well-positioned, service-oriented company."

Similarly, Beatrice Companies is currently transforming itself from a sprawling giant producing $11.4 billion in sales of everything from bras to sausages to a streamlined food-products company.

Colgate-Palmolive is another major corporation that is divesting itself of businesses that impede concentration on the company's core business strength. In a matter of a few months in 1986 Colgate agreed to sell off its baby-apparel business to Gerber and its athletic-shoe manufacturing, rice processing, and industrial-fabrics business to others. Whereas Colgate is selling, Gerber is buying. In the last few years Gerber has bought almost fifteen infant-clothing and juvenile-furniture companies. But *both* are advancing in the direction of concentration—Colgate to return to the concentration it knows best, Gerber to complement its core-strength concentration.

American Can Company's name is deceptive. It's an American company, but CEO Gerald (Jerry) Tsai, Jr., has skillfully shifted it away from its slow-growing, capital-intensive original business to a portfolio of service enterprises, including a highly profitable financial-services arm. Cans are now only a small part of American Can.

It's widely recognized that for most companies a small number—and often a very small number—of the products or services they offer pay most of the bills. Those products and the markets in which they flourish represent core strengths if concentrated on, while if the poor money-makers are eliminated, profits may well increase. I mentioned the president of a small training company who had decided to concentrate on sales seminars because the other thirty-four programs he offered didn't make money. Extremely large corporations operate the same if they're strong concentrators. Heinz, General Mills, Campbell Soup, Del Monte, and other major companies have achieved greater concentration by paring down their product lines. Westinghouse reduced the number of its refrigerator models to thirty from forty for the same reason.

And of course a core strength can be developed. If you don't have one, you can create one. For example, during a meeting with an executive of a huge Japanese manufacturing company I said, "I can remember when I was a kid and anything with 'Made in Japan' on it was flimsy junk."

The man laughed. "It was junk," he said. "Not anymore. We learned."

Attackers of Vulnerability (AVs)

If the enemy leaves a door open, you must rush in.

Sun Tzu

In 900, the *Tactica* of Emperor Leo the Wise of Constantinople admitted that the direct, head-on charge of mounted Franks armed with broadsword and lance could be terrifying and formidable. But, Leo added, the Franks were also disorganized and very poorly disciplined. To beat them, all you needed to do was to throw them into confusion by suddenly falling on their rear and flank—"a thing easy to accomplish." Strong in front, they were weak elsewhere.

Attackers of vulnerability—AVs—adjust operations in order to concentrate on taking advantage of each foe's peculiar weakness. And all foes have them. The opportunity of defeating the enemy is always provided by the enemy himself, said Sun Tzu. All nations, all armies, all corporations, and all people have a certain point of vulnerability, usually more than one, often many. The corporation doesn't exist, nor ever will, that is unbeatable everywhere. It's not good business to try to be. It's simply not profitable. All that any company will try to keep secure are the areas it considers key. Every other area will be only moderately defended or completely undefended.

Look at the leading firms in any industry at all—Exxon, General Motors, IBM, Du Pont, General Electric, etc. In every instance, their competitors demonstrate daily that even the very largest of firms cannot have it all their own way. Laws preventing restraint of trade aside, while the largest company could create serious problems for a much smaller competitor, it doesn't do so unless that competitor attacks what it holds important. And even if the giant does contest a market, it doesn't always win. For example, since between them the giants Coke and Pepsi hold the country's soft-drink market in their corporate hands, it would seem that all avenues are closed to potential competitors. Yet small cut-rate home-delivery bottlers are to be found almost everywhere—and along comes Jolt. Introduced in 1986, Jolt is for the person who has wearied of low this, low that, and particularly low sugar, low caffeine. In keeping with the product's motto "All the sugar and twice the caffeine," Jolt is loaded with 100 percent natural sugar and 98.3 percent of the maximum caffeine allowed by the Food and Drug Administration. Says C. J. Rapp, president of the firm, which is headquartered in Rochester, New York, "Who said soft drinks are supposed to be a health product?"

AV corporations—attackers of vulnerability—are thoroughly oriented to exploiting the competitor's weaknesses. They are competitor obsessed in a positive sense. In chapter 3 we watched Alexander the Great pour all his available troops through the one hole King Darius of Persia had foolishly allowed to open in his line. If you're an AV, you try to do the same. If a competitor is weak in marketing, the AV will tend to market heavily; if it underemphasizes advertising, if possible the AV will advertise more forcefully. AVs concentrate force at the weak gap in the competition's defense.

Sun Tzu says that wherever you find the opponent's main strength, very close to it you will also discover his chief weakness. To discover his vulnerabilities, look to the area of his strengths. Thinking along similar lines, Clausewitz contends that all your enemies have a *"centrum gravitatis,"* a CENTER OF GRAVITY on which for them everything depends—a core of power, a main strength. Clausewitz adds that if you can direct your blow there and defeat him, he will collapse, his main strength gone. Identifying and exploiting the enemy's center of gravity is, to Clausewitz, the key to victory. It's also the key element of some of the major battles of business warfare. Procter & Gamble is the sixteenth-largest company in America, whose products—Tide, Crest, and Pampers among others—are an everyday part of our lives. P&G has reigned supreme in the household-products industry for so long and is so widely regarded as the nation's best marketer that in the minds of some analysts it was considered invulnerable to attack.

Originally the strategy of its competitors was to try to understand and then emulate P&G's "magic marketing system" as best they could. Now that has changed, and its competitors have begun successfully to attack P&G's vulnerability. They are finding that P&G's chief weakness is located very close to its main strength, just as Sun Tzu suggested twenty-four hundred years ago. In Clausewitz's terms, P&G's "center of gravity" is its marketing system. And this is precisely what competitors are attacking. During the Battle of Blenheim (1704), the Duke of Marlborough's plan of attack was simple and bold. Marlborough chose to attack the enemy's strongest flank because it would be least expected. Marlborough feared nothing and no one and was willing to attack even the strongest of opponents. People with the same kind of nerves of steel are found in business warfare, and some of them are assaulting P&G.

Kimberly-Clark's chairman and CEO, Darwin Smith, has promised to "continue making life miserable for Procter & Gamble." The P&G vulnerability Kimberly-Clark has chosen to attack as it pits its diapers against P&Gs Pampers in P&Gs traditional marketing sequence—enter a market slowly, conduct extensive marketing tests to understand consumer preferences in detail, develop and test a superior product, and then, when everything is ready, launch a huge advertising program. It's the slowness in this cautious approach that Kimberly-Clark is attacking. For example, both Procter & Gamble and Kimberly-Clark began testing diapers with refastenable tabs at approximately the same time. P&G took the extra time to see that its product was exactly right, while Kimberly-Clark rushed in and captured sales. "I engage," said Napoleon, "and after that, I see what to do." That's what Kimberly-Clark did, too, but what P&G has trouble doing.

Tambrands Inc., makers of Maxithins, and another attacker of P&Gs center of gravity, developed and marketed its product in half the time P&G would have needed. When P&G's Duncan Hines division set out to develop new cookies, it took the traditional P&G approach of extensive consumer taste testing. The time taken gave Nabisco, Keebler, and Frito-Lay the opportunity to rush in with their own versions before P&G even reached national marketing.

Just as some of P&G's competitors are attacking slowness, so "cloners"—makers of low-cost imitations of IBM personal computers —are attacking IBM slowness when compared with cloners. All it takes to clone successfully (before a more advanced product is introduced at a lower price) is a small, quicksilver operation housed even in a garage, the kind of operation that can move faster than giants like IBM.

Lever Brothers is concentrating on P&G's price vulnerability, and Colgate-Palmolive is going after P&G's deemphasis of "faddish" products. P&G, offering fine products, usually charges a higher price for them. Lever Brothers captured a share of the fabric-softener market by introducing Snuggle at about 15 percent below P&G's price for Downy. Considering liquid detergents "faddish," P&G didn't push them; Lever Brothers did, and its Wisk captured (dare we say "whisked away"?) nearly a 40 percent market share.

Shortly after Black & Decker bought General Electric's small-appliance business, GE competitors, sensing a sudden vulnerability, moved quickly. Sunbeam increased its advertising budget by a staggering 400 percent to forty-two million dollars in hopes of taking GE's place as the best-known small-appliance maker. Norelco began marketing a line of irons to compete directly with GE's.

The sudden vulnerability that caused the furious, ferocious action had to do with a name change. As part of the acquisition, Black & Decker must put its own name on the GE products it bought—one of the biggest brand conversions in business history. Black & Decker must convince buyers, particularly women, that it not only makes a quality power tool but a fine toaster oven, too. If Norelco, Hamilton

Beach, Sunbeam, and others can persuade women that Black & Decker products belong in the basement and not in the kitchen, they will win this battle. What's at stake? A share of GE's half-a-billion-dollar small-appliance sales.

Discount brokerages like Schwab & Company seized on and exploited the vulnerability of retail brokerages—frills. All the discounters do is take orders to buy or sell. They don't give advice to the customer, and they don't employ commissioned sales people. No frills.

YOU CONCENTRATE; THE OPPONENT CONCENTRATES; YOU CONCENTRATE AGAIN

Concentration isn't something you do just once, then sit back and wait for your return on investment to rise. In World War II the French thought they could throw up the Maginot Line and relax. The Germans came in with their *blitzkrieg* (the essence of which is concentration), and France fell in short order. Business war, unlike military war, never ends. It's an endless chain of your action, the competitor's reaction, your concentration, and his counterconcentration. The sequence goes on forever. Every time the competitive sides are drawn, each side must reconstitute its focus—counterconcentration, counter-counterconcentration, counter-counter-counterconcentration, etc. For example, one of the more ferocious competitions in the United States is the one for people's bottoms—the competition to fill the seats of fast-food restaurants. There are more than seventy-five thousand fast-food outlets in this country alone, and most analysts believe that's too many. The industry is saturated. A shakeout is predicted. Some of the smaller chains will go belly up because they will violate the principle of concentration and try to expand too quickly. In 1983, 1 Potato, 2 Inc. went public hoping to displace the fast-food hamburger with the baked potato. Within two years it was shutting down almost 20 percent of its restaurants. Its original strategy was to sell baked potato entrées in shopping malls, but its management violated the axiom of adjusting your ends to match your means and expanded too quickly into building its own outlets.

The main form of concentration the newer chains take is specialized foods—the taco salad, for example, or baked potatoes, low-cal hamburgers, etc. To counter, the larger chains attack the other's concentration. Wendy's, for example, now serves baked potatoes and a light low-calorie menu. Says Robert Barney, chairman of Wendy's International, Inc., "You can't just serve a hamburger anymore."

Another example of responding to the competitor's concentration with a change in your own concentration is in the gas-station business. Amoco, Shell, Mobil, Texaco, and the other majors no longer bother with the expense of maintaining a repair shop, expensive equipment, and skilled mechanics. Their stations are "pumpers." They pump gas and perhaps sell a few popular grocery items. Grease Monkey, Inc., of Denver then concentrated on *not* pumping gas at all but changing oil and lubricating cars. Period.

THE THREE-TO-ONE FORCE-RATIO RULE OF THUMB

> *Move upon your enemy in one mass on one line so that*
> *when brought to battle you shall outnumber him, and*
> *from such a direction that you compromise him.*
>
> Napoleon

Napoleon declared, "It is always the greater numbers that beat the lesser." This statement of the emperor and others of its ilk have been hustled up to support the notion that the big guy will always win, a thesis that must have Napoleon turning in his grave and that is proved ridiculous in every arena of human effort from playgrounds to boardrooms. Since Napoleon was nearly always outnumbered and yet nearly always won, the man clearly meant something else.

He continues, describing how whenever he found himself with inferior forces facing a large army—which again was most of the time—he would "fall like lightning on one of the enemy's wings" and rout it. In the confusion that this maneuver *always* produced in the opposing army he would quickly attack another point, "*but always with my whole force.*"

What Napoleon meant, then, was that the more powerful force at the decisive point will triumph over the side less powerful at that point. TO WIN THE CONCENTRATION BATTLE, YOU NEEDN'T BE MORE POWERFUL IN GENERAL THAN YOUR COMPETITOR, BUT YOU MUST BE MORE POWERFUL LOCALLY, AT THE CRUCIAL POINT. You may be inferior in total, but if you can amass and concentrate sufficient forces to reach a local superiority, you can win. In the beginning of this chapter we saw GM growing stronger and Ford weaker, noting that GM is now the largest corporation of all. Yet the fact is that in Europe Ford sells more cars than GM! You can be smaller overall, and yet, by concentrating better than the big guy in one market niche or one territory or on one product, etc., you can outcompete him. Even as you're reading this, medium-sized and small companies across the country are proving this is true.

It's generally accepted militarily that the attacker must maintain at least a three-to-one superiority at the decisive point and at the time of the attack. If you're doing the attacking, it's ordinarily wise to try to be three times stronger than your competition at the point of attack. The principle of concentration also applies to the defensive. If you're the defender, you will be equally wise to prevent the attacker from building up a three-to-one superiority at any point you consider decisive. Let's say your competitor suddenly attacks, going after your largest accounts. The three-to-one rule suggests you put three times the effort into keeping those accounts than the competitor puts into stealing them. That's good advice.

This three-to-one rule, however, is just a rule of thumb, nothing more. Remember it, use it as a general guideline, but don't think of it as chiseled in stone. For while there are any number of examples proving this rule, there are about as many exceptions that disprove it and demonstrate that:

- The attacker can win with less than three-to-one local superiority.

- The attacker can actually be *outnumbered* and win.

It isn't very hard to compile a list of battles in which the attacker won
without a three-to-one superiority. In one study of forty-two selected
famous military battles between 1805 and 1973, in *most* cases the
successful attacker was actually outnumbered by the defender! Putting
aside who was on the offensive, who on the defensive, in twenty-four
of these famous battles the victors were numerically inferior; in
eighteen of them, the victors had more men. There are myths of
business warfare. One is that large companies are necessarily defensive
oriented. That is not true at all. Most giants have become big, stay
big, and grow bigger because they take the offensive habitually.
Another myth is that a small company that goes head-to-head with a
giant will necessarily lose—the battle and his shirt. That's not true,
either. For example, the WD-40 Company offers but one product,
yet outcompetes giants Du Pont, 3M, and Pennzoil in the multipurpose lubricant business. WD-40 has a total staff of fifty-five people
worldwide.

The larger army should win, but often the smaller force wins.
Why? One reason—perhaps the biggest—is because it's a better concentrator. It's totally committed to the approach the military masters
tell us wins—an all-out assault on a NARROW PART of the field of
action, not a fragmented attack along a large part. WD-40 has the
winning formula going for it—the right product in the most appropriate markets that pits the company's strengths against its competitor's weaknesses. But even if it didn't have all those factors
working in its favor, it still might win. In every instance of warring
and winning management, there is a minimum size of force necessary
to win, but this minimum can be *quite* small and may be enhanced
through *equivalents*, often intangible, that serve to multiply the
strength of forces.

Size matters but can find its match in certain other factors that compensate for lack of size. We saw this in chapter 1 when Gideon chose a smaller number of expert troops over a huge rabble. Alexander's men were usually far fewer numerically but battle proven and had the highest confidence in their leader. Their morale was supreme.

Leadership is just one intangible equivalent that multiplies force. Frederick the Great's Prussia was a pipsqueak of a country. At times it was completely surrounded by and at war with a group of hostile nations with much larger armies. Frederick said that the best hope of the small force lay in its leader. If the little guy's leader was better than theirs, the little guy would win. Frederick was so confident of his own abilities as a leader that he thought being surrounded played into his hands, giving the opportunity to reach any of his opponents fast. Napoleon's enemies claimed that his mere presence on the battlefield was in itself equal to forty thousand men. If Napoleon was worth forty thousand men to the French, how many does the mere presence of Lee Iacocca add to Chrysler's side in car wars? How many men does Victor Kiam add to Remington?

Innovation is another equivalent, and staff capabilities is another. The army of Swedish king Gustavus Adolphus was usually much smaller than its enemy's but made up of better fighters. His staff never asked how big the enemy's army was, just *where* it was. Genghis Khan's Mongols were almost always outnumbered. One Mongol cavalryman was probably the equal of ten of the enemy's best.

And what is the force equivalent and dollar value of a good, practical idea? Military history demonstrates that the side whose strategic and tactical ideas are better will tend to gain the upper hand, will do it faster and at less cost than the other side. In other words, if your managers Harry, Sally, and Bill are just better at thinking strategically —opportunistically—than their counterparts over in the competitor's conference rooms, you have the clear edge on them. While the three-to-one force ratio is a useful rule of thumb and generally is true, you may be outnumbered and still win. And if you do, says Clausewitz, you will automatically double the results. By beating "many with few," you will not only win the battle at hand, but you will make the competition realize that you're superior in ways it isn't. This realization, claims Clausewitz, will make them leery of tangling with you again. There's truth to what Clausewitz says. For example, in spite of its name, Giant Food Incorporated, Washington, D.C., is not a big supermarket chain. It's small. Yet through good local concentration and strong leadership it has routed far bigger competitors, including Pantry Pride, Lucky Stores, and American Stores. Not wanting to face Giant again, those chains left the area.

If possible, if you are on the offensive, be three times as strong as your competitors *at the decisive point*.

When on the defensive, try to keep the competitors from becoming three times more powerful at that point.

If outnumbered, exploit any intangible equivalents that will multiply your size, among them your own leadership.

TO CONCENTRATE, HAVE COURAGE

*When all is said and done the greatest quality required
in a commander is 'decision'.*

General Montgomery

You cannot possibly see, keep, defend, gain, or have *everything*; for this reason alone you must take risks. If you were consciously to set out to prepare a strategy of self-defeat, the first thing you would do is to overextend yourself. You would try to manufacture thirty products moderately well instead of fifteen extremely well and seven, or even one, better than any other company on the globe. You would open two hundred retail stores now instead of inching your way up to that figure. Peter Drucker's observation that concentration is the most often violated element of sound management is true.

To concentrate takes guts. Concentration and courage are constant companions. Every time you concentrate in business war, you're taking a gamble. Knight-Ridder Newspapers put seven years and twenty-six million dollars into developing its videotext Viewtron. Videotext is designed to connect consumers to travel information, financial data, home banking, shopping information, etc. Consumers didn't want it. Viewtron was junked. In car wars, GM is gambling that its concentration on plant-production innovations will pay off; Italian automakers are concentrating on sleek cars; VW is betting on higher-priced *and* supercheap models; Ford is going for small cars and joint ventures with Mazda; Chrysler is going for subcompacts and joint ventures with foreign manufacturers, etc. Some concentrators will be proved strategic geniuses; others will lose. Concentrators have the moxsie to declare their priorities clearly and boldly: "*This* matters; we will win *here*; the hell with the incidental." This is exactly what IBM did when it sacrificed other possible products and markets and sank five billion dollars in its System 360 series. Dun & Bradstreet showed the same kind of courage when, over the short span of eighteen months, it sold off its book-publishing and television properties while investing $1.8 billion in thirty-three acquisitions toward a concentration in the information industry. The timid never concentrate.

A VIRTUE OF WEAKNESS

I was too weak to defend, so I attacked.
Robert E. Lee

It's axiomatic that if you're too weak to attack the whole, you should attack a part. Being weaker than your competition in all but one or two areas can really be a blessing in disguise, *forcing* you to concentrate along those lines.

As your company grows larger and more complex, your concentration difficulties increase simply because you're now stronger in many areas. For example, if you market a single product in a specific geographic area, it is relatively easy to concentrate. But if you expand and offer a number of products in a number of areas, you suddenly face an increased variety of opportunities. You cannot concentrate on all of them or even many of them. Which will you choose? If you grow even larger, into a diversified corporation, you discover more that could be done and find choosing from among the options even more difficult.

WD-40 Company, mentioned earlier, is one corporation that's resisted the temptation to venture away from the company's *only* product—WD-40, the multipurpose lubricant that for thirty years consumers have used to lubricate squeaky wheels, clean rusty parts, and unlock sticky door locks. WD-40 has an annual sales of fifty-seven million dollars. Although small, it has never been displaced as the king of the many-use lubricants, even though giants DuPont, Pennzoil, and 3M have all tried and failed. WD-40 has a very fine product and stays with that product, layering on new uses for it. For example, traditionally WD-40 is seen as a man's product. The company is trying to show women that it's perfect for them, too. Also, according to the company's president, Jack S. Barry, the firm has always tried to find new uses in "various funny places"—mortuary and veterinary supply houses, for example, where WD-40 can be used to clean supplies and equipment. Barry's idea is to find two cans of WD-40 in every home—one in the garage, one under the kitchen sink. Sticking to one product is the basic strategy of this unusual company.

Research on extremely competent individuals reveals what managers knew all along—the best workers can do many things well. Some corporations and armies are like that—and that creates a concentration problem. If you can do many different things, you try to do them all, and all of them well, and that is simply impossible. It's always worthwhile, whenever you see too much happening, too many unrelated opportunities being pursued, and too many disconnected projects, to return to the one or two things you do best of all. There is usually one best strategy that exploits production and profit possibilities best for you.

CONCENTRATE OUT FROM YOUR
CENTER OF STRENGTH

*It is better to lose a province than split the forces with
which one seeks victory.*

Frederick the Great

"An army will encounter great difficulty operating far from its own turf"—this is another important lesson on concentration taught by the military masters.

Frederick the Great put it this way: "All wars carried far from the frontiers of those who undertake them have less success than those fought within reach of one's own country." Many corporations are learning the same lesson the hard way.

The Chicago *Tribune* said it succinctly when it entitled an article describing the failures of conglomeration "Conglomerates Going to Pieces." According to a study conducted by *Business Week*, consultants McKinsey & Company, and Standard & Poor's Computstat, of the forty-three most innovative and best-managed American companies cited in *In Search of Excellence*, at least fourteen—almost one-third—faltered within just two years because they couldn't adapt to basic change in the market or they drifted away from the business they knew best. Now these are not just any companies but the best-managed firms in the United States; certainly they were aware that a good measure of their success came from staying close to home in businesses they knew best; yet still they lost their concentration. It's that easy to do.

Even respected firms can encounter serious difficulty integrating disconnected businesses. Sears, American Express, Merrill Lynch, and Citicorp are encountering problems blending banking, financing, investing, real estate, and insurance businesses. Giant Exxon, the second-largest company of all, finally had to sell its office-products business after spending ten years and making a five-hundred-million-dollar investment. And General Mills had to sell off its toy and fashion divisions to concentrate more successfully on its consumer food products, restaurants, and specialty retailing.

J. C. Penney is the nation's third largest retailer. It lost its concentration when it made two mistakes: first, it deemphasized the more moderately priced clothes for which it was known and shifted to more expensive, trendy designer apparel; second, it left its terrain—small-town rural America—and opened stores in large metropolitan areas. After three years of declining profits and sluggish sales Penney decided to return to its original concentration. The less expensive clothes were returned to the racks, and many of the metropolitan area stores were closed. Said David F. Miller, Penney's president, "What we're doing is going back . . . where we can be the number one in town."

It would be suicidal madness never to venture out from your own corporate frontiers. When markets show signs of becoming saturated and growth slows, corporate leaders must venture out. For example, for most of its almost seventy years of existence the huge Japanese firm Matsushita has concentrated on home appliances and consumer electronics. Starting in 1983, it began shifting some of its concentration into office equipment, factory automation, and semiconductors. Says T. Fukahara, Matsushita's general manager of corporate planning, "In order to maintain double-digit growth, we've got to diversify."

Diversification into other areas is a hedge against setbacks in any one. General Motors compensates for auto-business cycles through its financial-services subsidiary. It would be unreasonable to speak of a firm that was not interested in diversification at one time or another. For example, after more than a century in the photographic business Kodak cannot afford to concentrate there solely any longer. In a short eighteen months in the mid-seventies it acquired five entirely new subsidiaries and moved aggressively into the floppy-disk and video markets and created units to produce biologic materials and electronic parts. Pointing out that an innovation does not last forever, Kodak president Kay R. Whitmore said: "We've come out of an environment where we were the single world leader, we had a technology that nobody else could really match, and we were able to dominate that field. . . . The world doesn't allow companies to do that anymore, so we've got to change, and that's a very hard lesson to learn."

The secret of successful diversification or any enlarging of concentration lies in how close the company stays to its original concentration or how careful it is to gain the know-how to operate in the new frontier before venturing into it. If ever a military master found himself needing to lengthen his line of concentration, he would extend from his center of strength, and even then he would never pull too far from the center. He would be very reluctant to abandon his center. In his comprehensive study of diversified companies, Richard Rumelt found that the best of them illustrate the same principle. Rumelt writes, "These companies have strategies of entering only those businesses that build on, draw strength from, and enlarge some central strength or competence. While such firms frequently develop new products, they are loath to invest in areas that are unfamiliar to management."

One strategy for the corporation venturing into an unrelated field through acquisition is to buy a small company before going big time. Japanese companies often buy little firms first because they provide a learning experience in how to concentrate in that business and at the same time minimize the risks because of the relatively small monetary investment.

Service stations that attach convenience stores to their "pumpers" grow from their center of concentration. Their extended concentration is reasonable because neither job—pumping nor selling—requires highly skilled personnel. Now these companies are adding delicatessens to further attract the one-stop buyer. This doesn't violate the principle of concentration.

The principle of extending from a strong center is implicit in those companies that venture out into far-flung fields but fields requiring the same set of skills and competencies that brought success in the company's original business. Executive search and recruitment may seem remote from accounting and auditing, but the Big Eight accounting firms that have diversified into executive placement have found the shift a natural. In essence, the skills of satisfying the corporate consumer were already present.

MATCH CONCENTRATION WITH OBJECTIVE

You can see the close concentration between this principle and the principle of maintaining your objective discussed in chapter 3. Once you determine what your objective is, then you can concentrate your power to achieve it; but if you're vague on what your objective is, then concentration to gain it is impossible.

OTHER EXAMPLES OF CONCENTRATION

There is hardly an aspect of corporate life that wouldn't benefit from a heavy dosage of active concentration. Here are a few examples:

In advertising. Consumers' memories are as narrow as they are short. People tend to remember just one thing from an ad—one concept, one claim. Concentrating your message on one theme is far more effective than changing your message.

In sales. Concentrate on exploiting your "unique selling proposition"—the qualities that distinguish your brand, your distribution, your product, your service, etc. If you're a salesperson, because you have more freedom of action than most anyone else in the business world, you have to be a better concentrator than anyone else. Concentrate on actually selling, and above all else, concentrate on closing. Simply by asking for the order your sales will increase between 20 and 40 percent.

Corporate strategy. A company's strategy requires concentration. It should be formed around a clear and ordinarily simple core idea of how the firm will exploit opportunities or beat competitors and respond to threats. If managers can't state their core corporate strategy in less than three sentences, the company probably isn't concentrating strategically.

Decision making. Concentrate on the most important decisions. If you're a manager, you're not being paid to make all the decisions, just the ones that—if they go wrong—would cost the company the most or that—when they go right—will save or make the company the most.

Leadership. Concentrate on leading, bold actions and being great at the critical moments.

Problems and opportunities. Typically, corporate managers are trained in how to solve problems. But most have never been taught to concentrate on the skill of finding—finding problems that need to be solved or, even more importantly, finding opportunities.

Small and mid-size companies. For best results these companies will concentrate on being the best in narrow market niches, being unique, or establishing a brand identity.

INTENSITY OF CONCENTRATION

There is concentration, and there is CONCENTRATION. Napoleon was the greatest concentrator of all time. After studying all the masters of the military art and comparing his methods with theirs, he came to the same conclusion. He concentrated better—and more intensely. His was concentration in a major mode—direct, powerful, even brutal—to shock, to overwhelm, to pour on the pressure until the enemy was crushed and completely destroyed. Every corporate concentration has a level of intensity attached to it. Tentative, low-powered, pussyfooting concentration has been shown time and again to bring meager results. Napoleonic high-intensity concentration brings victory. So whenever you decide to concentrate, do so with the highest possible intensity, or why bother concentrating at all?

THE ETERNAL QUESTION

When any project is put before you, ask yourself, "Is this consistent with good concentration?"

CHAPTER FIVE

Taking the Offensive and Maintaining Mobility

I am your king. You are a Frenchman.
There is the enemy. Charge!

Henry IV, Battle of Ivry, 1590

ON THE PLAINS OF MARATHON, 491 B.C.

In 491 B.C., a Persian invasion army of one hundred thousand men under Datis landed by armada approximately twenty-five miles north of the Greek city of Athens. The Athenian army, numbering ten thousand, wanted to avoid battling the Persians in the streets of Athens and so marched north to the plain of Marathon, where the Persians awaited them.

An Athenian war council was held on the slope of a mountain overlooking the plain. The question was whether they should do battle with the Persians, whose encampment lay below them. At the time, the Athenian army's management system was headed by a war ruler who held command for one year. His council included ten generals, one each from the ten tribes making up the Athenian army. Actual tactical command of the army was rotated daily, each of the ten generals being in charge every tenth day. A vote was taken. Five generals voted to fight; five not to.

One of the generals in favor of an attack—and an immediate one,

156

fast and furious—was Miltiades, one of the greatest of all military masters. A tough, decisive, confident man, Miltiades was head of the Chersonese tribe. Having been forced for a time to submit to King Darius and to serve in his Persian army, Miltiades was convinced that the Athenians held the upper hand despite being outnumbered ten to one. The other members of the war council probably expected such an aggressive attitude from Miltiades. On one occasion while serving in the Persian army, Miltiades and other indentured Greek commanders were told by Darius to hold a bridge across the Danube that Darius planned to use in the event he needed a fast escape from the enemy forces in front of him. When Darius had taken his army across the river, Miltiades boldly proposed to those with him that they destroy the bridge, leaving Darius no avenue of escape.

The factors Miltiades raised at the war council to defend his vote for quick attack probably included these:

- Because the Persians had so many more men, they would never expect the Greeks to attack. A sudden, audacious assault would catch the Persians napping—particularly their cavalry, which Miltiades had seen firsthand took hours to prepare to ride.

- The Persians lacked unity, but the Athenian ten thousand would fight as one man. The Persian army was hardly Persian at all. It comprised units of the many nations Persia had conquered. Their units didn't even speak the same language.

- Under heavy attack, the Persian army would break, every man for himself; the Athenian soldier would fight to the end. The Persians happened to have chosen for the battlefield the one place that in itself gave the Athenians a psychological edge. Marathon was sacred ground to the Greeks, the specific site where, according to their legends, their ancestors had fought off invaders in the past.

- The Athenians had intangible factors that offset the Persians' apparent tactical advantages. The bulk of the Persian infantry consisted of archers; the Athenians lacked even one archer. The Persians had one thousand cavalry, the Athenians none.

Discipline, leadership, and confidence would offset those factors, Miltiades must have argued—and speed. Above all, speed.

After Miltiades had presented his arguments, the deciding vote fell to Callimachus, the war ruler. He voted for taking the offensive. The Athenians would attack. Miltiades would assume the position of commander of the day for as many days as victory required.

On the afternoon of September 21, the bright sun reflecting off their armor, the ten thousand Athenians marched onto the plain of Marathon. Miltiades took his position at the front center of the line of troops. One mile away, directly across the plain, the Persian line stretched from horizon to horizon. The quiet small talk in the ranks stopped, a trumpet was sounded, and the Athenian line pressed forward; the Persians waited in place. Miltiades' main objective was to reach the Persians before their cavalry could mount, form up, and come out after him, before their unit commanders could coordinate their orders, and before their long bows took many casualties. His solution was simple, but it completely broke with the tradition that said infantry advances at a slow, methodical pace. Miltiades ordered a dead run for one mile. His troops were extremely fit, and he had no fear that they would be out of breath at the crucial moment of impact. He was right. His army reached the Persian line ready to fight.

The battle lasted all afternoon. The Persian cavalry never fully got into action. The Athenian troops broke around the flanks and cut down the Persian archers and infantry with their swords. Weary after hours of vicious fighting, the Athenians still maintained the momentum of the attack. Suddenly, in the early evening, the will of the supposedly invincible Persian army snapped, just as Miltiades had predicted. All semblance of organization and discipline quickly disappeared, and the Persians made a mad dash for the boats. The Athenians pursued, smelling victory, and fought on the beaches until the bulk of the Persian forces had clambered into their boats and had set sail and disappeared from sight. Sixty-four hundred Persians fell in the battle and 192 Athenians, a disproportionate ratio rarely, if ever, matched in history.

Realizing that the Persian boats would head straight for undefended Athens, Miltiades gave his men no rest but led them in a forced night march back across country to their city. In the morning, when the boats sailed into the Athens harbor, the Persian soldiers saw the troops that had beaten them the day before standing in perfect order, their lances raised, waiting for them. With that, the Persians turned the boats around and set sail for Asia.

PRINCIPLES

Hit first! Hit hard! Keep on hitting!
Sir John Fisher

This chapter discusses how two principles of war apply to business—the principle of offensive action and one closely connected with it, the principle of mobility.

THE PRINCIPLE OF OFFENSIVE ACTION states simply that victory can't be achieved unless at some point you take the offensive. As in baseball, no matter how strong your defense is, you won't win unless you manage to squeeze that one more run across the plate. Militarily, offensive operations are undertaken to destroy enemy forces, secure vital terrain, destroy the will of the opponent to continue fighting, deceive the opponent and divert his attention, or to gain intelligence. In business warfare you take the offensive to advance into your competitor's market segments or geographic terrain; to take away accounts, especially key accounts; or to reach a market or introduce a new product or innovation or price before your competition does.

THE PRINCIPLE OF MOBILITY, also called the principle of maneuver, advocates flexible movement to achieve a position of advantage over the competition. In his wonderfully titled *Military Rhapsody* (1779), military theoretician Maj. Gen. Henry Lloyd (1725–83) stated that "it may be established as an axiom that the army which moves and marches with the greatest velocity must, from that circumstance alone, finally prevail." In short, you not only need to take the offensive to win, but once taking it, you must move with speed.

MILITARY RESEARCH HAS SHOWN THAT THE COMBAT POWER OF THE ATTACKER IS AT LEAST 10 PERCENT GREATER THAN THAT OF THE DEFENDER. In corporate car war, the combat power advantage of the attackers producing small automobiles exceeded 10 percent. Volkswagens first appeared in the United States in 1949. The ten years it took GM to produce the first American counterpart was more than ample time for huge American losses in small-car sales.

The advantages to the company taking the offensive are many. The first with a product, for example, can establish a reputation that followers may have great difficulty displacing, can gain profits first, can benefit first from the experience curve, can set the standards, can establish solid relationships with distributors and retailers, etc. Companies that were the first to take the offensive have often remained the leaders in their industry year after year.

THE OFFENSE-VERSUS-DEFENSE CONTROVERSY

In war, the only sure defense is offense, and the efficiency of the offense depends on the warlike souls of those conducting it.

George S. Patton, Jr., *War as I Knew It*, 1947

Attack or defend. These are the only courses of action open to you in business warfare. Clausewitz wrote, "The defensive form of war is in itself stronger than the offensive." Yet more recent military thinkers have called that ridiculous and have completely reversed it: "The offensive is the stronger form of war." Well, which is it to be? The answer depends on what you're in the battle of business for—to win or not to be beaten.

The lessons of the military masters are that IF YOUR OBJECTIVE IS TO PREVENT YOUR COMPETITOR FROM BEATING YOU, REMAIN ON THE DEFENSIVE. BUT IF YOU WISH TO OUTDISTANCE THE COMPETITION, YOU MUST AT SOME POINT TAKE THE OFFENSIVE. Defense is a tool for forestalling defeat; it is not a tool of competitive superiority. "Ability to defeat the enemy means taking the offensive," wrote Sun Tzu. Even Clausewitz admitted that the defensive is less preferable to the offensive when he added only a few short lines after his famous statement on the defensive being the stronger of the two forms: "We must make use of it [the defensive] as long as our weakness compels us to do so," and "We must give up that [defensive] form as soon as we feel strong enough to." Napoleon didn't even feel that staying on the defensive would postpone defeat all that well. He wrote: "It is an axiom in strategy that he who remains behind his entrenchments is beaten; experience and theory are at one on this point."

An Active Defense Is Always Preferable to a Passive Defense

No manager and no corporation can be always on the offensive or continually on the defensive. Which you choose at any time depends on your inclination and the situation. Business war is an endless alternation of offense and defense. John Smale, CEO of Procter & Gamble, has described P&G's challenge of maintaining leadership in its established business and developing its position in new fields.

> By the time we came into the 1980s, we had developed major positions in categories we had entered earlier, so we felt we were again ready to move into some new areas. Since 1980, we acquired Crush, Norwich Eaton . . . and the Ben Hill Griffin Company in Florida. We subsequently developed and launched Citrus Hill Orange Juice. We entered the shelf-stable cookie field with Duncan Hines cookies. . . . So we moved into some broad new areas of business.

Smale is really describing the endless offensive-defensive sequence that companies, like armies, should follow: advance, secure yourself; advance, secure yourself again, advance again. The human quality required here is the patience not to move into another business too quickly. Companies don't usually sprout full grown; they inch up the vine slowly.

The secret is never simply to be satisfied with the defensive but to use it to serve the offensive. As a matter of fact, one of your units may be put on the defensive so that money can be put into another unit's offensive. Many people wrongly believe that "defense is defense"—that there is just one type of defense. In fact, there are two: one is good; one isn't. They're completely different. PASSIVE DEFENSE is securing yourself against the opponent's eventual thrust as well as you can, then digging in, keeping your eyes peeled, and waiting. ACTIVE DEFENSE is also called inelegantly but accurately "offensive defense." This is defense for the purpose of counterattacking and going on the offensive as soon as you reasonably can. With an active defense, even when you're on the defensive overall, you launch small offensive attacks and local forays. You stay active. When Kimberly-Clark's chairman and CEO Darwin Smith says he's going to "continue making life miserable for Procter & Gamble," he's expressing the real attitude of active defensive.

The active defense is ALWAYS preferable to the passive. The only defense that's ever worthwhile, productive, and promising is an active one. There has never been a military master in all of history who didn't oppose passivity in defense. To put it bluntly, no one in his or her right mind should ever take a passive-defensive stance—yet people do; yet whole corporations do.

One theme of this book is that all of warfare, including business warfare, is "psychological." Victory or defeat occur first in the mind, then on the battlefield. The psychological difference between taking the offensive and staying on the defensive is immense. Taking the offensive *immediately* invigorates and increases self-confidence; the main emotion of defense is fear. The side that has the initiative in warfare has a definite intangible advantage over the side that stays on the defensive, waiting to react to the opponent. You have *felt* this yourself. War has been called an "impassioned drama." It's much more impassioned and positive when you're taking the offensive. Even meetings at which you're discussing a new offensive against the competition "feel" far better than meetings for the purpose of figuring out how to cope with their offensive. Taking some offensive actions even if you're forced to be on the defensive overall is not only good warfare and good business but excellent psychology, too.

I saw this in business warfare some years ago when I consulted with Chrysler Corporation during its toughest times. In the days before Lee Iacocca arrived, some Chrysler managers would spend most of their time not doing anything that would help the situation. A defensive doom dominated the company. I remember walking into one meeting in which the topic of conversation was not business at all but who was going to be sacked next. After leaving that meeting, I stopped in at another. In that one, two hours were spent estimating just how much money overall Chrysler was losing each week, each day, each hour.

Iacocca's secret was the offensive-defensive stance he took immediately. Even though the company was on the defensive, its request for a federal loan guarantee was an offensive move. Immediately you could sense a complete change in the morale of Chrysler management and staff. While on the defensive, they suddenly had the positive confidence of being on the offensive.

The Main Advantages of the Offense Over the Defense

What are the advantages of the offensive over the defensive? The following are six of them. Though not valid in *every* instance, most of the time they are:

- Morale is generally higher on the attacker's side, as we've seen. Ardant Du Picq studied the reactions of defenders under attack. All the defenders have to do is remain calm and keep a steady aim. But they don't. The assault worries the defenders. They misfire or don't fire at all. They disperse; they run. On the other hand, the attacker's morale is raised. Clausewitz writes, "On no account should we overlook the moral effect of a rapid, running assault. It hardens the advancing soldier against danger, while the stationary soldier loses his presence of mind."

- The side defending a large area and long line must disperse its forces; the attacker can concentrate his.

- The proactive attacker has the initiative; the reactive defender must wait. Thus, the attacker is always one step ahead; the defender, one behind.

- The defender is in the dark as to where and when the attacker will strike. The attacker has surprise in his favor.

- The defender wastes many of his forces; the attacker uses almost all of his. The attacker will hit with concentrated force at only one or two points in the defender's line. The defender's forces not at those points will not even see action.

- The attacker can lose battles most of the time and still win the war; the defender is in trouble if he loses one battle. If the attacker hits ten important places and succeeds in breaking through in only one place, he has won; to win, the defender must be victorious at each of the ten places.

The new Remington company is just one company that demonstrates the advantages of the offensive over the defensive in corporate war. One of the most recognizable faces in all of American business—indeed all of America—belongs to Victor Kiam, the man, as you know, who liked his Remington Micro Screen shaver so much that he bought the company. In 1979, Kiam purchased Remington products from Sperry Corporation by borrowing twenty-four million dollars against projected sales and company assets and putting up one million dollars of his own. In four short years Kiam turned a company that had lost thirty million dollars in the three years before the buyout into a winner with yearly sales in excess of one hundred million. Remarkably, Remington's U.S. market share more than doubled to 41 percent. Following an enlightened leader's approach of "first you prepare yourself, then you attack an enemy," Kiam's first moves were defensive. He reduced staff and closed low-producing plants. Following the war principle of economy of force (see chapter 8), his next move included tending to the product and personnel. One hundred percent quality standards were introduced, and incentive plans were created. Also, Kiam established a direct, hands-on approach to running the corporation. During the workday he leaves the office and spends time on the line. Ready for the offensive, Kiam took it: his offensive had two major interrelated thrusts—advertising and making a celebrity of Kiam himself. Even not necessarily sympathetic industry observers consider Kiam's personality in itself to be a key to the company's rise. Leaders always leave a personal mark on those they lead. But only in exceptional cases is the personality of the leader the principal offensive weapon of the company or army.

ATTACKING

Hard pressed on my right. My center is yielding.
Impossible to maneuver. Situation excellent.
I am attacking.

Ferdinand Foch at the Battle of the Marne

Attacks are intended to create in the opponent the conviction that he's beaten. There are just three ways of doing it. By attacking from the front, the side, or from behind. Thus, there are three types of attack—frontal, flank, and rear—and six basic maneuvers. They have been with us since time immemorial. They are:

> Penetrating the center
>
> Enveloping one flank
>
> Enveloping both flanks
>
> Attacking in oblique order
>
> Feigning withdrawal
>
> Attacking from a defensive position

Together they form your offensive-attack "kit bag," the repertoire of options you can use when taking the offensive. You don't have to limit yourself to any one of them, but you may use more than one or all. Alexander the Great, for example, made use of virtually all of these maneuvers, depending on the situation. Napoleon customarily used the attack from the defensive position but was known to apply others if he thought they would serve his purpose.

Like a running back in football who bounces off one defender and heads another way, you can try one type of attack; if it doesn't work, try another. At the Battle of Hastings in 1066, William of Normandy tried to penetrate the strong Saxon entrenchment. Unsuccessful, he bounced off to the feigned withdrawal maneuver, pretending to retreat, drawing the defenders out, then turning and attacking. At Vicksburg, Grant used a number of trial-and-error "experimental" attacks. To other people they looked like failures, but they were really Grant's method of seeing which kind of attack would work best.

This testing the waters to see what works is just the type of strategic and tactical flexibility that wins battles and wars. Any corporation that undergoes restructuring—RCA, for example—is doing precisely what William of Normandy and Grant did (but with far more media attention)—eliminating whatever is unprofitable and concentrating on the more profitable.

Penetrating the Center

IBM dethroned Apple in personal computers (PCs) through another PC—penetrating the center. IBM simply copied exactly what Apple had done to achieve success—then did it bigger, better, and more boldly.

The object of the offensive maneuver of penetrating the center is to punch an entryway into a point in the enemy's front. Probably the most ancient tactical maneuver, frontal attack became sophisticated when some leader purposely held back a reserve force that he then threw into the battle late and used to overpower the exhausted defenders. Single penetration is breaking through at one point; double penetration at two points. You can use it when the competition is extended over a wide area. Having penetrated, you can encircle the enemy from his rear or push forward into his territory. The danger to guard against is concentrating at the center to the extent that you weaken your flanks and are easy prey for a counterattack. For example, you can become so engrossed in outdoing a competitor in one area—Florida, we'll say—that you weaken yourself along the East Coast and another competitor picks up and beats you there.

Sears, like IBM, is not afraid to penetrate any company's center. Traditionally, Sears attacks powerful, deeply entrenched adversaries. One of its more recent ventures involves attempting to penetrate the center of the bank-card market with its Discover card.

The Cummins Engine Company of Columbus, Indiana, is the leading producer of engines for heavy-duty trucks in North America. Knowing full well that Japanese firms were planning to enter the American market emphasizing lower prices, Cummins launched a preemptive strike, slashing its prices as much as 30 percent on some engines—*before* the Japanese arrived. Cummins chairman Henry B. Schacht says, "We made the decision we're not going to let them in."

One of business warfare's small but noticeable "hot spots" is the war of overnight mail delivery. In mid-1985 United Parcel Service (UPS) completed a single penetration when it lowered its next-day delivery charge to $8.50. This was between three to five dollars less *per letter* than Federal Express, Emery Air Freight, Purolator, and other competitors. This had great impact on the post office ($10.75 per letter at the time), but it had much less of an effect on industry-leader Federal Express. Why? As a key competitive element, Federal Express and others offer on-call pickup service; UPS doesn't. UPS drivers will pick up during their rounds; otherwise, the letter must be taken to a UPS center. By not offering on-call pickups, UPS isn't able to reach the professional or the very small company. Unlike larger corporate clients who ship cartons most every day, the professional (lawyer, accountant, dentist, etc.) and little company ship UPS sporadically. If and when UPS solves this pickup problem, it will complete a double penetration.

The hard times that have hit the smokestack industries have created a ripple effect. One industry that has been hit is Big Eight accounting. Some of their clients have been absorbed by conglomerates. Forced to find new customers, Big Eight firms are actively searching for new markets. Price Waterhouse is venturing out to mainland China, a flanking move into an area relatively unoccupied by competitors. But Touche Ross is attempting a penetration of the center attack. It's going after Wall Street, the territory of Deloitte Haskins & Sells.

United Brands is going toe-to-toe with Dole Foods in the frozen-fruit-bars war. United's Chiquita Pops is essentially mirroring what market-leader Dole is doing with its Fruit & Cream—spending lavish amounts of money on network-television advertising aimed at women with children, offering extensive couponing, and expanding into multiflavored lines.

Enveloping One Flank

Flank attack is the essence of the whole history of war.
Field Marshal Alfred von Schlieffen, 1833–1913

Even without knowing it, the majority of corporations are following the tactics of Frederick the Great. Frederick kept things very simple. Whenever he saw that his enemy had drawn itself into position, he attacked it aggressively on the flanks. FLANKING IS MORE LIKELY TO SUCCEED IN BUSINESS WARFARE THAN ANY OTHER OFFENSIVE MANEUVER.

It's a lesson of war to secure your own flanks and rear and to attempt to turn those of your opponent. So successful is flanking that Sun Tzu wrote that the force that hangs on to the enemy's flanks will eventually succeed in killing his commander in chief. Which for our purposes would mean that by clinging to the other side's flanks you will outcompete them and force their CEOs into early retirement.

There is narrow flanking and wide flanking. Which of the two you're doing is determined by how close or far you keep yourself from the competition's product, market segment, or terrain. And there are flanks and *strategic* flanks. A strategic flank is one that, if you turn it, will place the opponent's whole force in jeopardy.

To outflank the enemy is to extend your own line of forces beyond one of its flanks. It consists of swinging part of your force to turn an extremity of the opponent's line. Swing it around more and you envelop or get around the enemy's flank.

The advantages of envelopment are obvious; the possible disadvantage lies in your having to weaken yourself somewhere so as to build up strength on your flank.

There is the Big Eight in accounting, and then there is the fast-growing Laventhol & Horwath, number nine. Says Kenneth I. Soloman, chairman of the national board of L&H, "No one can out-Big Eight the Big Eight. Chocolate is chocolate, and there is no substitute for it." Rather than penetrating the center of Big Eight firms (their center is their ability to perform a complete audit for a multinational behemoth), L&H chooses the unprotected flank (financial management and tax planning for smaller companies and not a formal audit.) Will L&H ever knock off one of the eight Goliaths? It's on the way. Within just the last six years alone L&H has quietly completed thirty-eight mergers.

Rather than targeting its advertising to traditional flyers, as its competitors were doing, TWA advertised to people who had never traveled to Europe. This flanking movement alone accounted for a profitable season.

More and more companies offering "me, too" products (products like aspirin that are virtually identical to the competition's product) in mature industries are asking their managers the question "Given our business, how can we capture more of the market?" Many of those managers are answering, "Let's flank 'em through repackaging." Repackaging is what it says: increasing your side's sales not by offering a product different from everyone else's but by shifting the attack to how the product is packaged. The consumer has grown weary of and doubts "new and improved" claims but will buy the old product if the package does something different. For example, Gillette had problems convincing anyone that its For Oily Hair Only shampoo was really different from existing products that were a dime a dozen. But when Gillette introduced Brush Plus into the two hundred-million-dollar-a-year shaving-cream market, it had a winner. What distinguishes Brush Plus is that it's a clever device that dispenses shaving cream through a brush.

Other companies, too, are flanking the competition through better package design. For example, d-Con doesn't even bother advertising its insecticide formula; instead, it hypes the dispenser—a large insecticide-filled felt-tip pen that allows you to draw pest-killing lines in places where spraying isn't possible.

Toothpaste pumps also flank through packaging. In just one year after reaching the market, these pumping devices captured 12 percent of the country's toothpaste business. Charles L. Kane, president of a product-development firm in New York, was understating when he said, "People knew for years that the tube was not a good package because it's messy and you can't get all the toothpaste out. But if somebody takes the same product and makes it easier to use, what seems like a small improvement becomes a big thing."

Savin, manufacturer of small copy machines, avoided a frontal attack on Xerox. Instead, it attacked where Xerox was weak, at the lower price range.

T. J. Rodgers, president of Cypress Semi-Conductor, has in his words stayed "out of the way of the Japanese steamroller" by using an attack on one flank. While the Japanese and American juggernauts emphasize large-run, mass-produced devices, Cypress avoids head-on direct-market confrontation by selling customized chips. While the world's integrated-circuit market is expected to grow 20 percent annually over the next few years, the niche Cypress's flanking has uncovered could grow up to 100 percent each year.

By mid-1985, just four years after Emerson Radio Company was losing its war in component stereo systems, radios, and phonographs, its profits leaped 46 percent to $13.3 million. Emerson's win is a story of perfect timing and strategic flanking. Emerson switched to selling television sets at the precise time when video games and home computers increased the demand for TV screens. It has established a position in color TVs that's twice that of Hitachi and Toshiba combined by doubling its advertising budget and by skillfully flanking the competition on price. Emerson prices its products higher than the no-name imports but lower than the bigger brand names.

Flanking by offering a lower price or a higher-priced product is a commonly used form of enveloping one flank. You can also flank by producing a smaller-sized or larger product or one with features different from the competitor's.

Enveloping Both Flanks

Double envelopment—the famous "pincers movement"—is turning both flanks and surrounding the opponent. The double envelopment is the military maneuver *par excellence*. To the enemy it means surrender or extinction. Cannae, where Hannibal with fifty thousand troops double enveloped the Roman consul Varro with seventy-six thousand, is the classic example—the most perfect tactical victory in history.

It was on August 2, 216 B.C., that Hannibal drew his force into a crescent-shaped formation with cavalry on each wing and infantry at the center. After beating the Roman cavalry, Hannibal patiently waited for the attack of the enemy's infantry at his center. For a time he simply stood fast while the Roman infantry pushed back his own foot soldiers, forcing a bulge in Hannibal's line. At the critical moment, just when it seemed the Romans would burst through the bulge, he ordered his infantry on the flanks to turn inward to encircle the Romans. The circle was completed when Hannibal's cavalry charged in from the flanks. All but ten thousand of the Romans were killed, the rest taken prisoner.

To repulse the eight-billion-dollar takeover bid by financier Carl Icahn, Phillips Petroleum Company launched a two-pronged counterattack: (1) it created a new preferred share, and (2) it announced a proposal that would saddle the firm with excessive debt if corporate control changed hands. In March 1985 Icahn withdrew his bid.

"Bracketing" the competition's offering is double flanking. You can do it by offering a prestige product at a higher price than his and one at a lower price. You pull off your own Cannae by squeezing him on price and product.

Most "cloners"—makers of inexpensive imitations of IBM personal computers—attack IBM on one flank: price alone. The idea is to compete in an increasingly price-sensitive marketplace by selling as close an IBM imitation as possible but also as cheaply as possible. In response, IBM, which, one industry observer said, "knows how to get tough," will attempt a single flanking envelopment—and possibly the pincerlike double envelopment. In summer 1986 IBM slashed the dealer prices of its personal computers for its first flanking move. Although closemouthed on this, IBM seems to be making design or software changes that will create problems for cloners, however clever they are in trying to duplicate IBM's technology. If it does, IBM will have completed the pincer.

Attacking in Oblique Order

When you attack in oblique order, like a bulldozer you press increasing force against one end of the opponent's line until it gives. You imbalance your force at that one area until the opponent's line ruptures. The attack in oblique order, although usually associated with Frederick at Leuthen (1757), is really as old as history itself.

Scripto/Tokai believes that it can get consumers to stop flicking their BICs. Scripto was bought in 1984 by the world's biggest producer of lighters, the Japanese Tokai Seiki Company. Tokai Seiki turns out approximately ninety million lighters every month. Scripto/Tokai is basing its offensive strategy against BIC on the notion that even buyers of disposable items, particularly lighters, want more choice. Until forced to by Scripto, BIC had never introduced a new model since it first entered the business. Strong in the center of the line—low price—BIC was weak at the wings, offering the consumer just one product, just one model. That's the wing Scripto is attacking like a bulldozer.

For the price-conscious consumer Scripto offers the inexpensive Mighty Match; for the medium-priced buyer Scripto suggests the silver-and-gold Ultra Lite model. And for the less price-conscious there is the fancier Electra, called that because it features an electric ignition that's not as messy as the old flint and spark-wheel lighter. The Electra is designed especially for women, whose chief complaint with lighters is that they're dirty and break fingernails. For those with a naughty sense of humor, Scripto has a line of lighters imprinted with sayings like "Let's get foolish" and "Lack of sex causes insanity."

"Progressive domination" is a military term that could just as easily be a business term—the concept is the same. It means to gain supremacy progressively, gradually, and not in one fell swoop. Progressive domination is becoming more firmly and deeply entrenched in a market, expanding into other related areas of your business expertise, enhancing your products with new models, etc. Progressive domination is, like the attack in oblique order, bulldozing against your competition, but slowly.

Feigning Withdrawal

Make it appear you're retreating, sucker him into following you, then turn on him with a vengeance. The disadvantage lies only in the need to be timely in switching from retreat and turning to the attack. The feigned withdrawal has reappeared throughout history, bringing with it victory at some of the greatest military conflicts—Hastings, Austerlitz, Kadesh, and Salamanca.

The Mongols were the true experts in using this maneuver of luring the enemy in deep. At times they pulled their unsuspecting enemies after them for days on end, even slowing down or stopping to let the foe catch up. At the siege of the great Persian city of Meru, in the year 1221, Tolui, the most militarily gifted of Genghis Khan's sons, ordered a general assault against the well-defended city. The Mongols were driven back after sustaining considerable losses. The Persians, jubilant over their success, made the mistake of coming out after the Mongols, just as Tolui had expected. Once the Persians were on open ground where the Mongols could fight their style of warfare, the Persians were cut down in short order.

Procter & Gamble fell victim to this luring maneuver when an insignificant manufacturer of cleaning sprays purposely removed its product from stores when P&G was test marketing in the area. The small company waited until P&G had stocked the shelves with its product, then quickly slashed its own prices and just as quickly advertised extensively. Soon anyone who wanted a spray had bought one—the small company's. After months without sales, P&G abandoned the market, pulling its spray from stores.

Attacking from a Defensive Position

When an advancing enemy crosses water do not meet him at the water's edge. It is advantageous to allow half his force to cross and then strike.

Sun Tzu

The attack from a defensive position was Napoleon's favorite style of fighting. "The whole art of war," he said, "consists in a well-reasoned and circumspect defensive, followed by a rapid and audacious attack."

Take a strong, entrenched defensive position, let the competition exhaust themselves attacking; then, when it's sufficiently weary, attack. Until recently, Citicorp's position in Europe was a "circumspect defensive." For seven years following 1975 it didn't make one European acquisition. In 1982, as soon as competitors Wells Fargo and Marine Midland Banks began shutting their overseas offices, Citi suddenly shifted to an audacious offensive, acquiring banks in France, Spain, Italy, and Belgium; buying two major stockbrokers and an insurance broker in London; starting investment banking offices in fourteen European cities; and investing considerable sums in branch arrangements in West Germany and Great Britain. One European banking official called Citi the one bank in the world with the resources and guts to attempt anything so ambitious.

Napoleon used to listen to the sounds of artillery and musket fire, waiting until the sound told him the enemy was fully committed. Then and only then did he release the fury of his massed army. This is precisely what Wendy's did in the fast-food war. It let the smaller specialty chains settle in with their baked potatoes and diet hamburgers, then counterattacked by putting those specialty items on Wendy's menus.

Pittsburgh Plate Glass is an example of an offensive-defensive company. For the last five years PPG has been protecting itself from attack, selling $140 million in assets and cutting employment by 15 percent under what vice-president for chemicals Robert Duncan calls "a survival attitude." Now about to attack others, PPG is moving into a new high-tech business and plans to spend up to six hundred million dollars for acquisitions.

The marketing theory of "positioning" claims that the first company to the market and the first into the consumer's mind with a product will almost always win. This is not always true in marriage, where divorcées often find that the second mate is the better one; and it's not always true in business war, either. The first company usually has definite advantages, but there are corporations that, like Napoleon, launch very successful attacks from the defensive position. They bide their time; they wait and watch the product develop its markets; they learn from the mistakes of other companies. And then they attack.

Although leaning more toward "first to the field" status now, in the past Japanese corporations as a whole were rarely, if ever, first in a field. They let someone else absorb the high cost of research and development and original advertising, and only then swept in, improving the product and offering it more inexpensively.

Coke and Pepsi attacked from a defensive position in the diet-cola war. Royal Crown (RC) invented diet cola, introducing Diet Rite in 1962. Coke and Pepsi muscled their way over RC through massive advertising, the favorite weapon of firms that are offensive oriented.

Quaker Oats stayed back quietly in a defensive position after General Mills innovated the granola snack bar with its Nature Valley Brand in 1975. Soon after General Mills stepped on to the granola-bar battlefield, Quaker Oats surged past General Mills, capturing 45 percent of the market with its Chewy Granola Bars and Granola Dipps.

Many other companies are rarely first to the marketplace with a new product or innovation, preferring, like Napoleon, to attack from a defensive position. Some of these are among the best corporations in the world, including IBM, John Deere, Digital, and Hewlett-Packard. Each is an offensive-defensive company bent on benefiting from competitors' mistakes.

MAINTAINING MOBILITY, SPEED, AND MANEUVERABILITY

The art of war is simple enough. Find out where your enemy is. Get at him as soon as you can. Strike him as hard as you can, and keep moving on.

Ulysses S. Grant

The Race at the Battle of Salamanca

On July 16, 1812, the British army under Wellington found itself facing an equally large French army commanded by General Marmont on opposite sides of the Douro River. Devising an ingenious plan, on July 16 Marmont suddenly shifted to his right, crossed the Douro with two full divisions, and threatened the British flank. Wellington was equipped with as perceptive a *coup d'œil* as anyone ever had. He quickly gleaned the French intention and reacted immediately. He moved south, leaving two divisions behind. But as Wellington was marching, Marmont suddenly turned and countermarched swiftly, advancing directly at the British army. Wellington ordered his troops to quick march and led them toward a defensive position of safety. The French, closing ground, quick marched parallel to the British.

The race was on. For several miles on end the almost comical scene was of the two armies totaling almost one hundred thousand men racing at breakneck speed parallel to each other and so close at times that they heard each other's voices and enemy officers saluted one another:

> Hostile columns of infantry, only half-musket shot from each other, were marching impetuously toward a common goal, the officers on each side pointing "forward" with their swords, or touching their caps and waving their hands in courtesy. . . . At times the loud tones of command to hasten the march were heard passing from the front to the rear on both sides, and now and then the rush of French bullets came sweeping over the columns, whose violent pace was continually accelerated.

The British won, but barely, and eventually went on to win the Battle of Salamanca.

It's difficult to imagine a corporate manager who can read the story of the race at Salamanca without realizing how closely it parallels the competitive race. In business warfare, however cordially you might "wave" to the executives on the other side at conferences, you're out to beat them and they to beat you. The race is always on, and as this story illustrates, the side that has mobility going for it often wins.

Combat power is almost totally useless unless you can bring it to bear rapidly—at the right place and right time. That's the idea of MOBILITY. Napoleon's tactics were based on it—the sudden surge to the point of concentration—and so were the tactics of Frederick the Great, Miltiades, Gustavus Adolphus, Caesar, Sherman, Alexander the Great, Rommel, Patton, MacArthur—the list is endless. Almost without exception mobility was the ace in the hole of all the great military masters. Many of the most competitively superior corporations also recognize and exploit the potential of the principle of mobility. Another of business warfare's competitions within the competition is to move faster than your competitors.

One of the most Salamanca-like high-mobility actions in American business at the moment is the race to the top of the almost-two-billion-dollar high-priced superpremium ice-cream business. Pillsbury acquired Häagen-Dazs, the market leader, for sixty-six million dollars in 1983. In three years Häagen-Dazs's sales more than doubled to about $175 million. Then along came Frusen Glädjé. Bought out by Dart & Kraft in mid-1985, Frusen Glädjé tripled its market share in just one year. An attacker to the core, Frusen Glädjé has used hefty advertising spending to get its "enjoy the guilt" theme across. Feeling the offensive pressure, Häagen-Dazs is responding by quickly expanding its line to ice-cream bars, adding new flavors and increasing its own advertising budget. While the ice-cream business as a whole remains flat, sales of superpremiums are expected to continue to grow at double-digit rates into the 1990s.

In business warfare it's not true that the first to the field (the market, the innovation, etc.) *always* wins, but he often does. Inland Steel, Eastman Kodak, and First Boston are just three of many companies winning sales battles by installing in customers' offices product-listing terminals connected to their own terminals. Generally the effect is to encourage customers to buy your product and to lose some of their interest in the competitor's. These "channel systems" are truly offensive weapons. The Salamanca-like race is to be the first to install yours. As a *Fortune* article says, "The first system to appear in a market often becomes the runaway winner."

In business warfare, battle lines are never fixed. They move. Competitive forces are continually creating advances here and retreats there. All you have to do is read the business section of any decent-sized newspaper on any one day to realize that the battle lines of competitive business are forever shifting. The environment of competition is fluidity. Each strategy move by any of the competitors alters the whole competitive scene, making everything different from what it was even the day before. Even your own initiatives create ripples whose direction you cannot completely predict. The force that responds best is usually able to move faster, more flexibly, and with greater determination than the company that has trouble responding.

LyphoMed is one of the fastest-growing smaller companies in generic pharmaceuticals. It's involved in its own version of the Wellington-at-Salamanca race. In 1985 its profits rose 102 percent to $6.8 million, and its sales soared 116 percent to $68.3 million. Its ten-year aim is to reach a billion dollars in sales. One of LyphoMed's offensive goals is to compete with the giant pharmaceuticals in selling off-patent injectable drugs developed by those giants. Patent laws protect the creator for seventeen years. When the patent protection expires and the product goes "off patent," anyone can manufacture the product. LyphoMed believes it has at most two years to get a head start on other generic drug producers entering the competition and to catch some of the slower-moving giant patent holders.

Movement Is Our Way

War is a very simple thing, and the determining
characteristics are self confidence, speed and audacity.
None of these things can ever be perfect,
but they can be good.

George S. Patton

Someone once said that there is no national American physical type in the way there is a Russian type or Japanese or English. We're an amalgam of all the other types. But there clearly is an American psychology, and in war and business it's expressed in our disdain for niceties, nuances, and indirection. During World War II British and

American war strategists and higher tacticians didn't see eye to eye. The British were in favor of peripheral movements; our people wanted just to get across the damned English Channel to France and to strike into the heart of Nazi Germany. To beat around the bush is repugnant to Americans.

Americans prefer moving to standing still, action to inaction, decisions to deliberation, getting to the main objective to futzing around with incidentals, offense to defense, and mobility to passivity. In 1835, de Tocqueville said of us, "No sooner do you set foot upon American ground than you are stunned by a kind of tumult. Everything is in motion." This same preference for everything in motion is reflected in American traditions of warring. Throughout our military history there have been almost no slow battles. The men we hold up as our best stylists of war, like Patton, MacArthur, and Sherman, were masters of mobility. We think of aggressive, mobile fighting as both the best kind of warfare and *our* kind. This suits us perfectly for fighting mobile business warfare—wars of quick runs into the market and rapid responses to competitive opportunities and threats.

Study the Prototype

A half century ago a well-trained army division was capable of advancing between two to four kilometers per day. In 1939, the German army introduced "lightning war," *blitzkrieg.* In France, the German XIX Corps had its swiftest drive of ninety kilometers in one day. Against Russia, during the initial advance of the German Seventh Panzer Division, the rate of movement on the best day was 120 kilometers, thirty times faster than the trained divisions of old.

Blitzkrieg was refined by the Germans but certainly not invented by them. Sun Tzu described an "expanding torrent" in 500 B.C. Englishman Sir Basil Liddell Hart's books provided the German formulators of *blitzkrieg* with the basics of his version of the expanding torrent. Liddell Hart, in turn, had developed his theories of the deep-penetration method from the Mongols of the thirteenth century. Again illustrating that a good idea is not necessarily a new one, American general Douglas MacArthur was a student of Mongol Genghis Khan, borrowing from him a relentless insistence on mobility, quick offensive maneuvers, and avoiding the enemy's main forces to hit its lines of communication. "Were all of battle histories suddenly lost, except those of Genghis Khan," said MacArthur, "the soldier would still possess a mine of untold wealth."

The Mongol system was responsible for winning the largest contiguous empire in all of world history. To the Mongols, mobility was supreme. At one point, for example, they rode by horseback over fifty-five hundred miles, stopping to win a dozen major battles—in just two years. They sacrificed everything to movement, even the few minutes in the morning it takes to shave. They stopped their beards from growing by scarring their faces with knife cuts. It's no wonder the Russians gave the Mongols the name "Tatar," meaning "the people from hell."

Another term associated with the Mongols is "horde." Its original meaning is a Mongol tribe. It came to mean "vast numbers," because the foes beaten by the Mongols believed the secret of Mongol success was sheer numbers. If they had bothered to study the methods of their conquerors, they would have realized that the Mongols were almost always outnumbered, at times tremendously. But what they lacked in size they more than made up for in exacting training, rigid discipline, painstaking planning based on reliable intelligence, flexible organization, tough and skilled fighters, the use of surprise and deception, and above all else, mobility. The armies the Mongols confronted in Europe were unwieldy, immobile, and slow, virtually without maneuverability. They simply couldn't handle the Mongols, who rode on horses that were tiny but spunky and quick.

The principal elements of Mongol warfare are listed below. They can easily be applied to the offense-minded corporation.

- Attacking at a number of widely separated points to divide and confuse the enemy.

- Using deceptions to demoralize, terrorize, and unbalance the other side.

- Avoiding direct confrontation with a large force. When encountering an overwhelmingly larger force, the Mongols retreated, harassed, and lured the enemy out of his defensive position.

- Basing their philosophy of war strategically and tactically on ferocious attack. Employing fast-moving cavalry alone, all Mongols rode horses and rarely dismounted in battle.

- Refusing to stick to one tactical form—mixing things up but often pinning the enemy down with frontal attacks, then sweeping around to envelop the flanks.

- Reinforcing units with others. The Mongols advanced in a formation of five ranks. The first two ranks of cavalry were designed for heavy power and shock—using shields and wielding lances and hooks. Then came three ranks of light-armored expert bowmen who poured murderous fire over the ranks in front of them.

- Employing devastating waves of attack. Once a gap was opened in the enemy line of a walled city, the Mongols swarmed into it, hurling themselves into the defenders one after another day and night.

- Being ruthless in pursuit. When they gained an advantage— any advantage—they exploited it to the hilt.

- Leaving the enemy one open path of escape. Realizing that if you leave your opponent no choice but to fight, he will usually fight to the death, the Mongols left one avenue of escape open. When the enemy set foot on it, the Mongols swept down on him.

THE OFFENSIVE: MAKING YOUR OPPONENT DANCE TO YOUR TUNE

Always have as a goal to transform the war into an offensive on your part as soon as the occasion presents itself. All your maneuvers must lead toward this end.

Frederick the Great

Develop the Habit of Action

Throughout history this truth has never changed: in combat you must never be controlled by the enemy but should always try to control him. As long as you have him dancing to your tune, you have nothing to fear, but if he has you doing the dancing, you're in trouble.

By seizing the initiative through the attack, you carry the contest to the competition and seek decisions on terms of your choosing. All of which requires a strong capacity for . . .

Action. As mentioned earlier in the book, it's not coincidental that the first characteristic of well-managed companies cited in *In Search of Excellence* is "a bias for action," an attribute that seems to underpin the others. The best companies have the habit of getting things done. Similarly, in war the one factor that has contributed most to victory is simple human energy.

Many companies simply haven't fully developed the action habit. The same is true of most armies. "If we cast a glance at military history in general, we find so much the opposite of an incessant advance towards the aim, that *standing still* and *doing nothing* is quite plainly the *normal condition* of an army in the midst of war, *acting*, the *exception*. . . ."

What prevents the energetic forward movement of large bodies of people such as armies and work forces? Military history tells us three things do:

1. The fear of assuming responsibility.

2. Imperfections of judgment and perception, which are greater in intense action than under normal circumstances.

3. The greater allure of the defensive, since it is easier to take than the offensive.

All people can be divided into one of two groups: a minority are proactive, offensive-oriented individuals; the great majority are reactive and defense oriented. One of the best ways of picking up the pace of action in a corporation a notch or two is to put people with the action habit in charge. When the tank was invented, the question was raised who should command them? Some armies staffed them with infantry officers; but the wisest put cavalrymen in command, for the cavalryman was already conditioned to advance fast and to fight on the move. One can only wonder what the impact on corporation sales return would be if instead of filling home offices with traditional staff specialists, action-oriented salespeople—the business counterpart of cavalry—were put in charge.

I visit many firms, and whenever I do, I am struck by the tremendous differences between the tempo of one corporation's work force and another's. The employees of some of these corporations move in high tempo, as if they mean business. In other firms everybody's locked in slow motion, as if they're moving under water. Is this just a minor consideration? Not according to the military masters. Erwin Rommel, one of the geniuses of fighting mobile war, was admired even by the men who opposed him. Rommel felt that by just observing the continuous tempo of staff you could predict which side would win and which would be taught "a bitter lesson." Anything less than high tempo right from the beginning will eventually lose, he said.

Battlefield Tactics

The system of modern offensive battlefield tactics was developed by Gen. Erich Ludendorff, a deep thinker and German chief of staff. Because it was first applied in actual combat by General Oscar von Hutier in his offensive against Riga in September 1917, this tactical system has since been called "Hutier tactics." You can use Hutier tactics when launching your competitive offensive by emphasizing these factors:

Surprise. Know as much as you possibly can about the battlefield. Make use of product intelligence, market intelligence, and competitive intelligence. Know the consumer and the competitor's leaders. Keep your plans and build up from the competitor's attention. Your offensive should be sprung in full force. The competition should have no inkling of what you've concocted until you're already in the field.

Locating and hitting weak points. Avoid prolonged attacks where the competition is strong. Attack; if the competition's resistance is formidable, back off and attack elsewhere. Continue these probing attacks until you locate a "soft spot" and penetrate there. The *Schwerpunkt* is a German term meaning point of main effort. Find your *Schwerpunkt* and attack.

Exploiting success. Any and every unit of yours that penetrates must press the offensive straight ahead. Units leading the assault should not be restrained. They should not be told to stop after having reached X objective; they should be given no stopping point at all but told only to keep pushing straight forward regardless of delays to units on their flanks. If you break through in sales, push selling even more. If your advertising clicks, advertise harder. Don't have the lead unit wait for the others; have the others catch up. Think of your offensive as three distinct phases: infiltration that disorganizes the opponent, then the breakthrough, then the follow-through pursuit. Napoleon showed that those three phases should (and could) be one continuous sequence.

Maximizing fire support. Seeing your concentration, the competition will probably fire back, using weapons such as intense advertising campaigns, price cutting, etc. Rush reserves up to the front from the rear. Support the lead units. When attacking, all parts of the assault should be synchronized and orchestrated, for an "all arms" attack, just as an effective marketing campaign is tailored to specific market targets. Shock and overwhelm at a narrow point, but deeply and continuously. Break through at full throttle and keep pressing on relentlessly. Bring all your coordinated resources and units to bear and throw them into the attack: manufacturing, finance, marketing, advertising, sales—everything.

Maintaining momentum. Whoever is maintaining the momentum is winning. That's why every great military commander and corporate manager strives for continuous attack. When you have any of the competition's elements on the run, "kick 'em in the ass," as Patton said.

A Reasonable, Not Suicidal, Offensive

The spirit of the offensive must always be sensible. The absurdity of a maniacal offensive was demonstrated at the Somme during World War I. Troops were thrown forward madly, and sixty thousand British soldiers alone fell in the first three hours. Be hard driving and relentless after you launch your offensive but cautious, careful, and calculating before. Never attempt an offensive unless you're superior to your competitor in some way. "The offensive, always the offensive" is an attitude that nearly got the French defeated in World War I. "The offensive almost always but not if it's absurd" is a better one. Even though you can't win without taking to the offense, if you're up against a competitor who is more powerful than you in your business and your markets, for the time being at least you have no choice but to take a defensive stance.

The Duke of Marlborough broke from the customary defensive method of fighting of his day. He chose offensive-attack strategy only because he realized that two factors played in his favor and gave him a tremendous advantage: the flintlock and the bayonet— offensive weapons. He fought only because he had the wherewithal to win; otherwise, he would have avoided a fight.

Remember, adjust your ends to your means. If your means don't tell you "offensive," remain on the defensive; but if you have the means to take the offensive, take it as soon as possible.

When to Attack

Here's an age-old problem that most every general and capable manager has faced. It's balancing the time factor against the security factor. You're thinking, for example, that if you wait (for more information, to see what the competition does, etc.), you can make your side stronger and better prepared. But on the other hand, you're also thinking, *But if we don't move now, we may lose our chance.* Unfortunately, the general or manager who provides the definitive answer to the question "What should I do?" has not been born. Some have won by waiting for a while; others have been equally, even gloriously, victorious by going ahead, anyway, knowing full well they weren't completely ready. Hope you have a good intelligence system; hope you're smarter than your counterpart on the other side; pray you're luckier—that's about all that you can do.

You can also think of the future. If it holds out better things for you, stay on the defensive and wait for it. But if the future holds more promise for your competition, take to the offensive now. Recoton Corporation of Long Island quickly entered the stereo-TV-conversion business knowing that its product, F.R.E.D.—Friendly Recoton Entertainment Decoder—would have a short life span of two to three years. The company couldn't wait; it had to move quickly. F.R.E.D. created the sound of stereo TV without the consumer having to buy a stereo television. Whenever the Romans felt that a war with an enemy was inevitable, they declared one immediately, believing that to defer was to bring advantage to the other side, which would build up. They didn't wait for the inevitable. Neither did Recoton.

Another factor you should consider when you're weighing the advantage of waiting or attacking is the "personality" of your competitors.

Every corporation has a dominant personality. There are:

- Companies that will defend but not attack. If they're more powerful than you, go around them. Find a niche different from theirs. But if you have nothing to fear from their defense and they have what you want, attack.

- Corporations worn out by fighting. They're susceptible to attack.

- Firms managed by people with weak character. Strike suddenly; steal their morale from them.

- Companies that are full of fighters but disorganized. Each unit fights for itself. Attack and harassment work best.

- Corporations of well-organized fighters. Be very careful.

Finally, it's clear to me that managers and corporations that prefer mobile, offensive action are usually bold. If you're one of the bold, the advice of panzer general Heinz Guderian might be right for you: "When the situation is obscure, attack."

Attacking an Entrenched Competitor

If you have decided to attack an entrenched marketing opponent, Frederick the Great would advise:

- That you do it before he solidifies his position, if possible.

- That your success depends on knowing the strength and weakness of the entrenchment. Their weaknesses are generally of two types: (1) the entrenchment is not solidly enough supported, or (2) it is too wide for the number of troops assigned to it. If the first is the weakness of the entrenchment, attack where it is weakly supported. If the second is its weakness, fake an attack at one point (called "masking" the attack). When the competitor reinforces it, launch your real attack with the majority of your forces at the vacated spot.

- After you have captured the entrenchment, reinforce it with all your forces.

Any company that's introducing a fledgling product into a market characterized by strong brand loyalties is attacking an entrenched competitor. Folgers is a good example. When it went into the coffee market, it was aware that it would have to spend liberally if it was to overpower its entrenched competitors. It prepared the assault well in advance with its "We're bringing a mountain" to your community advertising. It realized, too, that light introductory advertising, like light gunfire, held little promise of breaking an entrenched defensive position. To overcome powerful brand loyalty, it would have to use heavy artillery—distributing free samples. Probably the strongest form of advertising, free samples are also the most costly. In all, Folgers gave away more than one million pounds of free coffee.

Launch a Series of Interlocking Offensives

Once under attack, the competition will usually not quit without a fight but will send reinforcements. This will "thicken" his line at those points you've attacked.

Facing thickened points, your managers will be tempted to shunt off and go around them. If everyone shunts off, you soon have a situation in which every manager is fighting solo. The same thing happened to the Germans in World War I.

In that war the French general Ferdinand Foch defeated the famous German general Ludendorff. Foch explained his winning strategy by pointing out that in all of the tactical details Ludendorff's plan was perfect. It could not possibly have been better. But Ludendorff had failed to realize that one attack, however perfect, doesn't win battles. To succeed, one attack has to be linked to a number of others. One attack mustn't be seen as the whole effort but as a single part of a larger whole. This, said Foch, "is what Ludendorff forgot."

Foch, then, emphasized *a series of offensives*, one "interlocked" with another, a sustained succession of rapid attacks at different points but all coordinated to serve the whole, the larger objective. When the Germans resisted at any one point, the French launched another strike at a different point, the first, second, and third coming shortly after the preceding one. The Germans were paralyzed.

The importance of continuous movement and of a series of co-ordinated offensives has been proved time and again in business warfare, too, particularly by the more powerful corporations.

Texas Instruments uses this same series-of-offensives approach. At any one time TI may be pursuing a dozen or more separate opportunities. Every opportunity is assigned its own full complement of staff, yet all are coordinated to the whole through a system of comparing each with respect to progress and results.

Advertising, too, is based on the notion of a series of offensives. In spite of the tremendous volume of advertising, one study revealed that only one of four women spontaneously recalled *any* advertising during the previous week for the product she had just bought. Yet sustained over years, that advertising helps her to make clear distinctions and form buyer preferences.

The advertising concept "synergism" and "accumulation" are just like the Fochian series-of-coordinated-offensives approach. The effect of one medium alone is not as powerful as the effect of two media working together. Each message builds on those preceding it. More than twenty years ago James O. Peckham demonstrated that the successful introduction of a new product requires a series of advertising offensives: introductory offers to the trade and consumers, enough advertising over a two-year period to produce a share of advertising about one and a half the share of sales you intend to attain and sufficient advertising in the following months to maintain the product's share "modestly ahead of its sales position."

Combining arms and corporate units leads to victory; using them separately, to defeat.

The Culminating Point

By "the culminating point" Clausewitz meant that point at which the momentum of the offensive shifts to the defender. Beyond that point "the scale turns . . . and the violence of the reverse is commonly much greater than was that of the forward push." The only sensible response, says Clausewitz, is to pick up the offensive again, perhaps in a new direction.

In business warfare each product has its own "culminating point."

The life cycle of every product follows the pattern of introduction, growth, competition, obsolescence, and termination. It's during the last stage of the pattern, termination or phaseout, that you either eliminate the product from your line or assume a new offensive in the form of improving the product or finding different uses and new markets.

For most of this century traditional drugstores had things their own way. They reached their culminating point when discount stores entered the health and beauty field; their problems were compounded when supermarkets attacked, too, also selling health and beauty products. Drugstore chains like Revco, Eckerd, and others made deep cuts in their prices and saw profit margins dwindle. Like an army whose offensive has waned, they must carry the offensive in a new direction—toward less competitive store locations, perhaps, and growth niches like all-night convenience stores whose customers are less price conscious.

Quaker Oats reached the culminating point with its rice cakes and is recasting the product to appeal to a broader market. Currently, only 4 percent of American households consume rice cakes. Quaker Oats is adjusting its advertising pitch to the 30–40 percent of homes that are fitness oriented and the estimated 10 percent of homes that at any one time house a dieter.

In addition to finding new markets, a few avenues for action at a weary product's culminating point include adding a feature, finding new uses for the product, and modifying packaging.

"YA GOTTA SPEND"

The notion that the loser suffers more casualties than the winning side is not always true. During the Civil War the victorious North suffered more than one hundred and forty thousand more casualties than the South, and in the First and Second world wars the Allies suffered about double the losses of the vanquished Axis.

If a company is to win, it must eventually take the offensive. It must eventually attack, penetrate, or outflank the rival. This costs money. . . . Ya gotta spend. For example, to exploit fully the advantages of a new product takes considerable initial investment in market development, staffing, facilities, etc. In most industries the top brand spends less on advertising relative to total sales than its competitors spend in proportion to their sales. You must outspend your competition in advertising for every point of market share you hope to win for a new product—at least in the short run.

You can succeed in any attack, every attack, if you can pay the price and are willing to. Particularly if the competitor is deeply entrenched, secure, and fighting in his terrain, any attack you launch will result in initially high losses on your side—you will have to pay, and perhaps dearly, for every can or bottle or item you sell.

For example, Schick's advertising expenditures in 1970 and 1971 were under 2 million dollars annually. When it went on the offensive to increase the sales of its "Flexamatic" electric shaver in 1972 and 1973, it spent $4.5 million and $5.2 million in advertising during those two years, according to industry sources. By late 1972 Schick's market share had doubled, from 8 percent to 16 percent.

Some executive once said, "Half of what I spend on advertising is wasted, but I don't know which half." And someone else said, "When you reach the point where you're running the company to conserve cash, it's not a question of whether you're going to fail but how soon."

At times you have to spend in spite of yourself. For example, when two industry powerhouses do battle head-to-head, other smaller, higher-cost competitors are forced to spend so as not to fall too far behind. Every time the Japanese enter a specifically targeted market, they spend money by losing it on each item sold. They price their product considerably lower than their competitors, hoping to gain in the long run what they lose in the short run. Competitors are very hard pressed to remain in the fight.

The question is: Does your company have the financial resources to wage an aggressive offensive war?

UNDERSTAND THE PSYCHOLOGY OF THREAT

After observing the responses of animals to threat, zoologists coined the term "critical reaction." The reaction of animals to a threat is determined by how near or far the threat is from the animal. When the threatener is beyond a particular distance, the animal will retreat. That's the "flight distance." But when the threatener is inside that particular distance, the animal will attack. This is the "critical distance." It varies from species to species, person to person, and from one company to another. It's highly individual. Some guys let people kick sand in their faces; others don't. It's the same with companies. In 1978, the Miller Brewing Company's top management stated publicly that Miller was going to be number one in the beer business. William K. Coors, chairman of Coors beer, was with August Busch II, head of Anheuser-Busch Companies, when the declaration was made. "I'll never forget the look on his face," recalls Coors. "He said, 'Over my dead body.' And he meant every word of it."

The "territorial imperative" is the same idea. Living creatures instinctively protect their territory. The farther away from the center of it you stay, the less interested in you the animal is. But step over the border and into his territory and the more aggressive he will become. Raccoons mark off their territory by urinating around it. Corporations are only slightly more subtle.

Some corporate managers, like some warriors, are continually walking the delicate, invisible line of demarcation between the opponent's flight distance and his critical distance. Others don't give a hoot and step right into the competitor's critical distance.

Coke and Pepsi are currently engaged in tough business warfare. For years Pepsi irritated Coke, but Coke restrained itself from stern reaction until Pepsi showed no respect for its critical distance and stepped right into it. Pepsi is a staunchly aggressive company that cares very little about observing anyone's territorial claims. Paralleling its fight with Coke is the fight of Pepsi-owned Taco Bell. Pepsi is attacking the entire fast-food hamburger business, including Mc-

Donald's and Burger King. And it's not worrying about critical distances. Realizing that the only way it's going to get more customers for itself is to take them away from someone else, Taco Bell is spending seventy million dollars on an advertising campaign that in part ridicules the hamburger. In one spot a man threatens to jump off a ledge rather than eat another burger.

McDonald's executives rarely make public speeches. So when its chairman makes a speech, its competitors should listen. In one recent address president Michael Quinlan blasted competitors who use comparative advertising. "If you have a marketing department and the best they can do is tear down Brand X, I'd get a new marketing department," he said. Powerful McDonald's is feeling annoyed around its critical distance.

In mature industries companies generally are careful that any offensive action they pursue will not seem to the others to be a disruptive threat. They don't want to violate the other firms' critical distance.

The "show of force" has a long tradition in warfare. Its purpose is to avoid entering the enemy's critical distance but to stay in the flight distance and yet demonstrate what power could be unleashed—to give the opponent the opportunity to surrender, or in the McDonald's example, to tell the competition to cut it out. On the morning of the Battle of Waterloo, Napoleon had his armies march before him in review—within eyeshot but out of range of the Belgians and British, whom he hoped to intimidate into abandoning their positions. Unfortunately for the emperor, who was literally about to meet his Waterloo, they didn't scare easy.

The "buffer zone" is a tangible expression of the defender's need for protection of his critical distance. It's the little space around the edges that provides companies some comfort. If you launch an offensive into an opponent's buffer zone, the defender takes it to mean that you fully intend to attack his critical distance—and he will often fight.

Price wars, as in the airline industry (often called "bloodbaths" by airline executives), are sparked off by some one company violating the critical distance of the other airlines by powering through the buffer zone of the "going rate" and charging right into the center of the other side's territory. Price wars are direct frontal attacks, and they often result in a "pinwheel strategy," each side attacking and trying to drive the competition back. One airline industry defense against the devastation of total price war is being made possible by a weapon that's both offensive and defensive—the computer. In past years, airline price wars were nationwide and automatically affected all seats. Now by computer monitoring of travel patterns and the number of seats being bought at low fares, airlines can target price reductions for particular cities and certain seats.

Advertising wars are less harmful than price wars. By increasing advertising in the industry, all companies may well benefit by an enhanced need for the industry's product. The Victor Kiam-Remington "I bought the company" advertising has been so effective that the entire industry has profited. A few years after Kiam started making the ads, the U.S. shaver market expanded by 50 percent.

Territorial issues of what is mine, what is yours are of immense importance in business warfare. Sun Tzu classifies "ground" into nine types and suggests tactical measures for each.

1. *Dispersive ground.* When your own territory is under attack. As any good corporate manager knows, the idea usually is to try to keep competitors off your ground. Try not to fight on your own territory; but if you're forced to, unify the determination of your army. Also, you may retaliate by fighting on *your competitors'* territory.

2. *Frontier ground.* When you make just a shallow penetration into enemy territory. Don't stop on the competitor's borders but push in, advises Sun Tzu.

3. *Key ground.* Ground that imparts great advantage to the side

that holds it. If the competitor firmly occupies it, don't attack. Go around to his rear or flank.

4. *Open ground.* Terrain that's equally accessible to both you and the competition. Concentrate your forces; don't allow them to become separated. Pay strict attention to defending your positions.

5. *Intersecting ground.* Ground that is surrounded by three other states. To gain control of it, form alliances. Value your allies dearly.

6. *Serious ground.* When you have penetrated deep into hostile territory and find getting back difficult, you're on serious ground. Reap as much as you can while you're there.

7. *Difficult ground* is rugged terrain that's hard to cross. Press on.

8. *Encircled ground.* Surrounded by competitors, you must resort to stratagems, surprises, ruses, and tricks of war and business.

9. *Death ground.* Any "ground" in which the army survives only by fighting for its life is death ground. Compete with the courage of desperation. Act speedily; don't delay your decisions. Let your personnel know the seriousness of the battle. "It is the nature of soldiers to fight to the death when there is no alternative." At least the best of them; the rest will bail out.

Some Lessons on Threat

• Corporate managers are possessive territorial animals. They'll fight harder on the defensive to protect their land (product niches, markets, reputation, etc.) than they will on the offensive to get your land. One method of taking some of the wind out of your competitor's offensive sail is to turn and attack what he holds dear—his territory. Robert Galvin, chairman of the Motorola Corporation, talks about "disciplining" Japanese firms whose price cutting in U.S. markets is hurting American firms. Like other American firms, Motorola is launching a counter-attack in Japanese markets. For example, in late November of 1986, Motorola agreed to enter into a joint venture with

Japan's Toshiba Corporation to manufacture semiconductors. The arrangement will enable Motorola to increase its access to the Japanese market and to enhance Motorola's knowledge of the extremely efficient manufacturing processes of Japanese memory chip manufacturers. Said Galvin, "As we have embedded ourselves stronger into their marketplace, we can have the ability to discipline their market practices." He added, "We can tweak their noses a little bit when they tweak our noses in this market."

- A threat to your rival's flight distance may make him retreat; one to his critical distance will arouse him to fight you.

- Aggressive types, whether nations, companies, or individuals, can't be appeased. But since they respond to force, they're susceptible to threats. If they strike you, strike back.

- The corporate leaders who will resist your attack to the bitter end are the ones that initially didn't want to fight you but whose deep defensive emotions were struck.

- If you don't particularly care if a competitor fights back, attack his critical distance. A few weeks before the Japanese attacked Pearl Harbor, they sent a peace proposal to Washington. Roosevelt, wishing to "maneuver them into a position of firing the first shot without allowing too much danger to ourselves," replied with a tough ten-point memorandum that the Japanese, not surprisingly, interpreted as an ultimatum.

- Before moving against a competitor, ask yourself, "What is his critical distance?" Every person's and every company's is different. What's his? If you think you know what his is and you intend to move into it, anyway, ask yourself when his counterattack will come, what form it will take, and how effective is it likely to be?

- Always be careful not to blunder stupidly into a competitor's territorial center—his "killing zone." The defender can ambush the assaulting force. Intelligence on the competition's expected countermoves can help, and so can holding back a reserve to suppress retaliation.

EXPLOITING SUCCESS THROUGH PURSUIT

*Next to victory, the act of pursuit is the most important
in war.*

Clausewitz

We'll say that your company has just made a successful attack and
has come out of the offensive the winner. What happens next?

In many firms nothing happens, or next to nothing positive,
virtually nothing that keeps the momentum of the attack going. In
the late 1700s Maj. Gen. Henry Lloyd complained that the British
had developed the bad habit of never pushing through an attack to
its most fruitful and decisive conclusion. "Our military institutions
exclude every idea of celerity," he said, adding that because of it
English victories were never really complete, never totally decisive.
The English attack petered out, the enemy went off to occupy some
hill or another, and the English had to start the whole battle process
over again.

On September 8, 1944, Company G of the American 115th
infantry division raced seven hundred yards in five minutes across
an open field under brutal German fire, shooting from the hip and
shouting the division battle cry as they charged. Most of the Germans
fled, leaving only a handful behind a hedge for cover. No more than
a mere fifty yards from total decisive victory and against a few
Germans, not one member of Company G advanced even one inch
during the next seven hours! What causes this slowdown when real
victory is so close?

It's the mental and motivational slump after intense action. Whole
work forces sometimes want to relax when they feel they've done the
job and are drunk with success and basking in a sense of well-being.
It's well documented that people tend to relax in the general vicinity
of a goal—when they're very close to reaching the goal or have
reached it or exceeded it. This phenomenon is widely known in
athletics. To counter it in track, runners who would otherwise relax

and slow down as they get closer to the finish line are taught again and again to run not *to* the tape but *through* it. "Run through the tape."

After victory a person feels secure and tends to forget about the competitor who's been beaten. Frederick advises you to think of the other side *immediately* after you've won. Put yourself in his place and ask, "What would I do if I were he right now? What project could I form?" In short, always feel nervous about the competition, particularly after you've won.

Pursuit is a subprinciple of war—and without doubt the most often violated of all. Pursuit means simply *don't* quit when you're ahead or you'll probably find you've suddenly fallen behind. Instead, continue the attack or, more boldly, launch a new action before even stopping to assess the results of the first. Contact once gained should not be lost.

Just one company that will benefit from the application of the pursuit principle is Adolph Coors Company, makers of Coors beer. Says a beverage industry analyst, "The Coors pattern is to enter a market with a big splash and in the first six months sales skyrocket. Then, suddenly, almost inexplicably, sales start to slip as the mystique wears off."

When a strategy is proving effective and the opportunity still exists, pour more into the strategy. Pursue. Loser companies are reluctant to put greater allocations into a program than budgeted; on the other hand, winners allocate more if the opportunity can be validated. "Without pursuit," wrote Clausewitz, "no victory can have a great effect." Exploit every gain fully. The Mongol light cavalry pursued the enemy so quickly and vigorously that they rode through the gates before they could be shut. Most of the best-managed companies pursue and exploit in the same way. Once they gain a foothold in a market, they expand their product lines to reach a wider segment of the market.

Advertising scheduling is just one small element of business warfare that depends on pursuit:

- Weekly exposures to advertising develop a faster rate of recall than the same number of exposures spread over time.

- If you don't continuously expose the consumer to advertising, he or she will forget.

- The more exposures, the slower the rate of consumer forgetting.

Frederick the Great of Prussia introduced innovations that made his cavalry into one of the most dynamic assemblages of horsemen who ever thundered across a parcel of earth. Historically, cavalries rallied to the rear. Prussian cavalry was trained to rally to the *front*: in other words, to regroup together after a charge *while still in pursuit*. Of the twenty-two significant battles Frederick fought, his superb cavalry alone accounted for at least fifteen wins. If corporations trained their managers and staff always to rally to the front instead of stepping out of action and taking a breather for a while, they might well win more battles, too.

MAINTAIN THE DEFENSIVE DESIGN

Frederick, a master of the offensive, designed everything for movement, creating horse artillery; using light, mobile howitzers; and relentlessly training his men to drill and maneuver with lightning quickness. Having once established these design features of the offensive, he never forgot to maintain them—to keep them in the act. History shows us how unusual he was. Cavalry is designed for movement; yet by the end of the fourteenth century the typical cavalry horse carried the weight of its rider plus 150 pounds of equipment and armor. The advantage of mobility—the whole damned original idea—was lost.

The moral? Don't weigh your people down with excess weight—reports, meetings, nit-picking policies, and overly cautious management. Lighten things up. Pursue.

CHAPTER SIX

Following the Course of Least Resistance

A soldier told Pelopidas, "We are fallen among the enemies." Said he, "How are we fallen among them more than they among us?"

Plutarch

HANNIBAL, 218 B.C.

At twenty-nine—an age that would find him a junior executive in most major corporations—Hannibal, the father of strategy, led an army of fifty thousand Carthaginian infantry and nine thousand cavalry on the famous march across the Pyrenees and the Alps. He emerged in northern Italy and defeated the great Roman commander Scipio's troops on the banks of the river Ticinus. Then, on December 26, 218 B.C., in a snowstorm he crushed two Roman armies along the river Trebia.

Two more Roman armies were raised and sent north to block Hannibal's path to Rome. The armies were under the command of the Roman consuls Flaminius and Geminus. Realizing that his opponents expected him to take the obvious roads to Rome, Hannibal decided instead to follow a shorter but more dangerous route through treacherous marshes and to come out on Flaminius's unprotected flank. "Normal soldiers always prefer the known to the unknown," writes B. H. Liddell Hart, the twentieth century's best-known theorist of war. But, he adds, Hannibal, like other great commanders, was "abnormal" and hence "chose to face the most hazardous *conditions* rather than the certainty of meeting his opponents in a position of their own choosing."

Pushing on, without sleep, wading waist deep across the flooded marsh for three nights and four days, Hannibal's army finally came out on dry land. In front of them they saw Flaminius's encamped army. Hannibal could have marched on Rome, but being a great battler, his immediate objective was not Rome but the Roman army.

Like any great leader, Hannibal had taken the trouble to study the man opposing him. He had learned that Flaminius was extremely proud and headstrong and easily insulted. So, to goad him into a foolish attack, Hannibal insulted him. He moved his troops defiantly in front of Flaminius and made camp just a short distance down the road from the Roman army. Without Flaminius knowing it, Hannibal had hidden his light and heavy infantry and cavalry so that they couldn't be seen. Advancing down the road, Flaminius's army passed the first of the troops Hannibal had hidden in a gully. A thick fog lay across the land that day and further hid the Carthaginians. Unsuspecting, the Romans marched on. Suddenly, the Carthaginians closest to the camp rose to their feet, cheered, and attacked. With that, all of Hannibal's forces immediately poured out of their hiding places. Shouting battle cries, they converged on the Romans from all sides and destroyed almost all of them, including Flaminius. Hannibal's victory that day was the most celebrated ambush in military history.

THE COURSE OF LEAST RESISTANCE IS
USUALLY AN INDIRECT ONE

*Everything which the enemy least expects will succeed
the best.*

Frederick the Great

Strategy and tactics can take two forms: (1) the direct approach
and (2) the indirect approach. The former consists of a direct advance to the enemy, culminating in a powerful frontal attack designed
to overpower him. The indirect approach involves coming at the
opponent from a roundabout direction that he's not totally prepared
to resist. Throughout history, most great generals have consistently
chosen the indirect approach, risking almost anything to catch the
enemy with his guard down. Like Hannibal, they have scaled moun-
tains, brought their armies through swamps, and suffered every
imaginable hardship rather than face an enemy that's ready and waiting. The best course for great commanders has usually proven to be
THE COURSE OF LEAST RESISTANCE. Following the course of least
resistance is also one of the most important elements of corporate
competitive superiority. ONE OF THE CHIEF COMPETITIONS WITHIN
THE COMPETITION OF BUSINESS WARFARE INVOLVES IDENTIFYING
AND EXPLOITING THE COMPETITION'S LINES OF LEAST RESIS-
TANCE.

B. H. Liddell Hart was probably the most influential twentieth-
century military strategist. He shaped his entire theory of war around
the advantages of the indirect approach. He wrote, "The history of
strategy is, fundamentally, a record of the application and evolution
of the indirect approach." If Liddell Hart isn't considered the fore-
most military strategist of our time, his friend, British major general
J. F. C. Fuller (1878–1966), is. Fuller challenged the notion that
the indirect approach was the end-all of strategy. He put his criticism
very simply. "The object is to defeat the enemy and if this can be
done by a direct approach so much the better."

They're both right, as your own common sense tells you. If Hannibal had felt that he could have beaten the Romans with a head-on
attack, he would have ordered his troops straight ahead, but realizing
that his chances of winning would improve if he went around, that's
what he did. If you're confident that you can beat your rival in direct
competition, that's how you'll compete. You'll put your gas station
or store right across the street from his. But if your better judgment
tells you your chances will improve if you avoid him, you'll figure
out some way of going around him. You'll try to find a higher
traffic corner than he's got, for instance.

Generally, though, both business and military history reveal the
advantages of the indirect approach over the direct. For example, in
his analysis of the misfires of corporate marketing campaigns, Thomas
L. Berg attributes the failures of many such unsuccessful marketing
efforts solely to the use of the direct method: "Marketing failures
have often resulted from head-on attacks against the entrenched
positions of stronger marketing rivals." Berg cites examples of
General Foods withdrawing a baby-food line after a futile direct
approach against Gerber and other brands, and Swanson and others
defeating GF's direct attack in the frozen-pie business. Corporate
strategy expert Michael E. Porter makes the same point: "Some
companies seem to view competitive moves as entirely a game of
brute force. . . . However, even sheer resources are often not enough
to insure the right outcome. . . ."

The book *Three Plus One Equals Billions* is the story of how the
Bendix Corporation attempted to buy Martin Marietta and was taken
over by Allied Corporation. It is an indictment of the direct approach
that often finds America's largest companies setting out to swallow
apparently for the sake of swallowing. During World War II only
two victories were achieved through the use of the direct approach
(El Alamein and Stalingrad); and in the three hundred campaigns
of the thirty major wars during the twenty-four hundred years before
1939 a decisive victory was achieved by the direct method a grand
total of six times.

Like corporations, armies should know better than to attempt direct attacks against a stronger opponent. But often they don't. In its war with Russia in 1904 Japan threw its armies directly at the Russians, wearing itself out in a bloody series of brutal battles that ended in a standoff. The Japanese considered themselves lucky to be able to make peace. An indirect approach would have won. If the Japanese generals had simply checked a map, they would have seen that the Achilles heel of the Russians was their depending entirely on just a couple of strips of steel, two railroad tracks—the Trans-Siberian Railroad, which brought their supplies and armaments to them in the field. Had the Japanese simply gone around the Russians and blown the tracks, they would have quickly brought the Russians to their knees.

In business warfare almost every corporation should be using and refining indirect approaches. The indirect approach is really the only course for any corporation that's not the market leader. Which means that the direct frontal attack is a viable option only to one or two companies in an industry. For everyone else the indirect approach of gaining sectors is best. One of the surest signs of corporate wisdom is to hear managers who have worn themselves out in a frontal attack against an arch competitor suddenly say, "Hell, let's go around them from now on."

How can a manager proceed who's interested in taking an indirect approach against the competitor's line of least resistance? You can use surprise, think originally, exploit your competitor's values and priorities, and you can create lines of least resistance through "tricks of war." These methods are described in the following sections.

Surprise Follows the Course of Least Resistance

War is the realm of the unexpected.

B. H. Liddell Hart

Hannibal and his brothers Mago and Hasdrubal, Genghis Khan and his strategist Subadai, and Confederates Robert E. Lee and Stonewall

Jackson were three of the greatest teams of fighters in history. Coincidentally, all three exploited another of the principles of war—surprise. Jackson's maxim is the most memorable: "Mystify, mislead, surprise your enemy," he said. Clausewitz said that to one degree or another surprise is without exception the foundation of *all* military undertakings.

Surprise has been called "any commander's greatest tactical weapon." It was certainly Hannibal's. But surprise operates at the strategic level, too, as we learned all too well on the morning of December 7, 1941. Why are surprises that work effective? Because they follow a course of least resistance. The most stunning surprises are the result of a novel, creative idea. Creativity consists of connecting two or more heretofore unrelated ideas or things. Napoleon was as creative a general as there has ever been. He once gained a decisive surprise on the enemy by connecting manure and cannon. The emperor's object was to move up under the cover of night and to surprise the opponent by being in a completely new position in the morning. "Impossible," he was told, for the only passable route was across a rocky mountain road. The clanking of the wheels of the artillery carriages would certainly eliminate any possibility of surprise. Napoleon looked at the road, the guns, and the horses, then ordered the road filled with horse droppings to deaden the sound of the wheels. It worked.

The crafty Mongols once won a decisive battle by creatively shodding their horses. The Mongols drew the enemy Tangut cavalry onto the ice-covered Yellow River. The Tangut mounts slid out of control, unable to put up resistance, while the Mongol horses were steady and surefooted, for the Mongols had wrapped their horses' hooves in felt.

A move that surprises your rival often results from your own sheer audacity. For example, at the Battle of Denain, 1712, Marshal de Villars purposely committed the cardinal sin of exposing his flank and crossing a river right in front of the enemy's entrenched army—and against an experienced and skilled enemy commander, Prince

Eugene. The prince found it unbelievable that de Villars could possibly do something so stupid and for hours sat watching without comprehending, choosing to believe it was merely a small detachment. All Eugene had to do was march his troops straight ahead and he would have won easily. Instead, he ordered them to break for dinner and himself sat down to eat. He realized what really was happening only when he saw the entire French army, now on his side of the river, forming for an attack. Eugene lost his entrenchment.

In warfare, a preliminary bombardment might weaken the enemy's lines but eliminates any advantage you might have gained by surprise. So there are two schools of thought. One advocates bombardment; the other forgoes it and chooses surprise. The same kind of disagreement exists in business warfare. Some companies go for intensive marketing surveys even though by doing so they give away any hope of sneaking up on the competition. Other companies choose the element of surprise even if it means entering a market without a clear idea of what the demand might be. "Ready, fire, aim" is their motto.

Abandoning Traditional Ways of Thinking Often Uncovers a Course of Least Resistance

Asking yourself, "*Why* does it have to be that way? Who says?" time and again can often result in catching your competitors with their guard down. For example, Napoleon considered Turenne (1611–75) the greatest French general of all. In Turenne's day it was an unwritten rule of decorum that gentlemen didn't wage war in winter. They rested. Possessing one of those minds that's always asking, "Who says?" Turenne decided that winning a war was more important than having others think him a gentleman. During one winter campaign he gave the appearance of settling down into winter quarters, deploying his troops to separate encampments. When the enemy saw this, they did the same. Turenne suddenly regrouped his army, swept south through a snowstorm, and defeated his foes with a quick, decisive attack.

More than a century later Frederick the Great found himself also facing gentlemanly, winter-resting enemies. Having studied Turenne, he borrowed his winter ploy and also won. During our Revolution, captured Americans would swear allegiance to the king. But they were no gentlemen. Freed, they would immediately take up arms. The English—gentlemen, of course—were totally exasperated. Industries, too, have codes of "gentlemanly" conduct that are unnecessarily restrictive. Companies that ask themselves, "Who says?" usually gain the initiative on competitors. One afternoon a broker friend of mine stopped in to see me. "Just look at him," he said. There was no one else in the room but the two of us. "Who him?" I asked, looking around. "Him," the broker said sharply, pointing to a financial newspaper lying on my desk. The him was Charles Schwab, whose picture was on a Schwab ad. When Schwab started his discount brokerage, he followed a course of least resistance by breaking traditions right and left, including the one that seemed to upset my friend the most. Schwab advertised *himself*. "Egomaniac," said my friend.

Your Competitor's Values and Priorities
Open Courses of Least Resistance

A course of least resistance is open to you if you pursue a line of action that your competition is unprepared to counter. It may be because of his philosophy, his values. For example, when accounting firms first began moving into business consulting, the accountant's code of professional conduct that discouraged advertising for business was applied to consulting, too. On the other hand, the consulting firms that hadn't come from an accounting background were not impeded by the no-advertising tradition. Handcuffed by their own values, accountants-turned-consultants were unable to resist. After some years the no-advertising code was abandoned.

As we saw in chapter 4 of *Waging Business Warfare*, Lever Brothers and Colgate-Palmolive exploited a line of attack that they knew full well more powerful Procter & Gamble wouldn't resist because of its values. Traditionally P&G doesn't like to make products it considers "faddish." Its two competitors didn't particularly care if the product was faddish or not as long as it was a good product and turned a profit.

You can tell the size of an army and the direction in which it's headed by the location and density of the dust it raises as it moves. If you watch your competitor's dust and take a different route, you can often come upon a course of least resistance. For example, if you see that your competitor is diversifying heavily into financial services (and who isn't?) he may be unable to resist your powerful push of four new product models. He may want to counter you but simply not have the cash to do so.

A course of least resistance is afforded to you if you offer a product that doesn't fit well with the competitor's or potential competitor's portfolio. One of the most dangerous things a company can do is tamper with its successful image. You can turn this to your advantage by offering a product or service that's in conflict with that image.

Courses of least resistance are as plentiful as blades of grass. You can take advantage of little resistance if you invest in something that doesn't fit the competitor's strategic plans; or if you pursue an opportunity that the competition overlooks; or if you attack the terrain of a competitor that shows little inclination to compete aggressively or aim at customers he neglects. Differentiation, segmentation, and niching are each marketing concepts designed to allow you to identify and pursue the competition's line of least resistance.

Creating Lines of Least Resistance: Stratagems, Not Strategies

In war the skin of a fox is at times as necessary as that of a lion, for cunning may succeed where force fails.

Frederick the Great

All that military strategy consisted of until about 1800 were stratagems, *ruses de guerre*, "tricks of war" that a general used to deceive his adversary and reduce his resistance. *Ruses de guerre* are part and parcel of business warfare, too.

If you can get me to believe that it's just a waste of my time and money to bid on a contract because you've got it sewn up and I don't bid on it, you have just increased your chances of winning the battle for that contract. I have no chance at all. What have you done? You haven't *discovered* a line of least resistance; you have *created* one.

Now this kind of thing is done by corporate managers every day in business—probably upwards of five or ten million times each day across the globe. It isn't how much power you actually possess that dominates the rival's decision to oppose you or not to but how much power he feels you have and will use. In business warfare, as in military wars, each side is trying to control the opponent's thoughts. When clothing buyers applaud designer creations at fashion shows, they make it a point not to clap loudest at the design they like best simply because they don't want the competitor's buyers to know what they intend to order.

Byzantine emperor Leo the Wise was known to increase his troops' morale by making them believe they were part of a more powerful army than they actually were. He would write highly detailed accounts of great victories in other theaters of war and have them circulated among his troops. What the soldiers didn't know was that Leo had fabricated them completely, from beginning to end. So convincing were some of these reports that even to this day historians are not quite certain which battles really happened and which are Leo's fictional handiwork.

Sun Tzu wrote that true excellence in warring consists of breaking the enemy's resistance without fighting. Corporate strategist Bruce Henderson comes to the same conclusion. In his book *Henderson on Corporate Strategy*, he contends that a businessman likes to think of himself as completely logical when in reality the critical element is his emotional biases in comparison with the competitor's emotional biases. He goes on to say that two points are worth emphasizing. One, that a company's management must convince each competitor to stop short of maximum effort to gain profits and customers; two, that convincing them to do so depends on emotional and intuitive factors rather than on analysis or deduction.

The surest way of preventing your competition from venturing into a "terrain" you consider yours—your market segment, your customers, your geographic area, etc.—is to make him *believe* you will use more force against him than he can handle and to make him *feel* in his gut that he must hold back rather than go ahead. The difficulty is making him believe it and feel it without your incurring the expense of actually doing it.

Military masters have gone to great lengths to reduce their opponent's desire to resist, including constructing dummy tanks, planes, and bridges; raising dust by dragging mats to give the appearance of a large approaching force; starting rumors; marching infantry one way in the clear light of day, sneaking them back at night, marching them back the next day to give the appearance of strength; using double agents, etc. Belisarius lit a long, stretched-out chain of campfires to deceive the larger Gothic army into believing that he had more men than they.

GUERRILLA WARFARE

There were giants on the earth in those days.
Genesis 6:4

S. J. Prais, the English researcher and author, recently sent me one of his books, *The Evolution of Giant Firms in Britain*, along with a note: "Hope you enjoy this book and find it useful." Since it was a totally unexpected gift, I felt obliged to read it. The first paragraph caught my attention: "The growth of large industrial enterprises of a size undreamt of in earlier centuries is a modern phenomenon, the progress of which is traced before our eyes almost day to day. The cumulative effect of this evolutionary process has been the creation of a new world of giant businesses. . . ."

After noting that this trend toward the concentration of power in a relatively few corporate hands also occurred in the United States, Prais describes its three stages of evolution. During stage one, before the mid 1800s, the concentration of industrial power was at a low level. During the one hundred years between 1850 and 1950—the second stage—large corporations on the average grew slightly less rapidly than small firms. But during the third stage—from 1950 to the present—"factors systematically favouring a relatively faster rate of growth by large firms have become dominant . . . to make for an unprecedented rate of increase in concentration, to which no limit can be seen at present."

Few would deny that concentration of power is occurring in some industries on an enormous scale and with breakneck speed. As of February 1987, six megacarriers control almost 80 percent of the U.S. airline industry—Texas Air, United, American, Delta, Northwest, and TWA. Texas Air, which may fly as many as one of every four U.S. air travelers in 1987, includes Continental, People Express, Eastern, and New York Air. Combined, three other carriers control just 11 percent of market share—Pan Am, USAir, and Piedmont. Together, the remaining nineteen jet carriers account for less than 10 percent of U.S. air traffic. In the movie and television business, the conditions are right for a potentially massive concentration in the hands of just three or four enterprises by 1995. On November 6, 1985, John Welch, chairman and CEO of General Electric, paid a half-hour social visit to Thornton F. Bradshaw, chairman of RCA. Welch brought up the idea of merging the two companies and creating one company that he described later "can compete with anyone, anywhere, in every market we serve." Six weeks later, GE agreed to buy RCA for more than six billion dollars—the largest nonoil merger ever.

Corporations like GE and the others in which phenomenal power is concentrated can fight almost any kind of warfare they wish to.

Everybody else (certainly the more than 90 percent of all the corporations in the United States that employ one hundred or fewer people) have no option open to them but one—to fight guerrilla warfare.

No style of warfare follows the course of the opponent's least resistance as consistently as guerrilla warfare. The guerrilla's strategy and tactics are founded entirely on attacking points of low resistance. Mao Tse-tung codified the fundamental principles of guerrillaism. They are so simple that Mao communicated them in just twenty words. "THE ENEMY ADVANCES, WE RETREAT; THE ENEMY CAMPS, WE HARASS; THE ENEMY TIRES, WE ATTACK; THE ENEMY RETREATS, WE PURSUE."

Those few words contain the basic approach followed by any guerrilla fighting anywhere on Earth and at any time in history. The guerrilla is the master of the course of least resistance.

In the military sense guerrilla warfare is waged by small bands of irregulars who live off the land and are armed only with light weapons. They don't have heavy artillery and aren't lavishly supplied. Their opponent is. Guerrillas fight against regular armies who not only wear uniforms but are generally well supplied and much better equipped with light arms and heavy armaments.

"Guerrilla" is Spanish for "little war." It came into use during the Peninsular War (1808–14) in Spain, when Spanish irregulars helped the British by harassing the French to the tune of costing them one hundred men each day for seven years. Americans have a long tradition of waging guerrilla warfare. Rogers's Rangers used guerrilla tactics during the French and Indian Wars (1754–63). That George Washington did so, too, is attested to by the fact that Mao Tse-tung studied Washington's methods. During the Civil War, both the North and South used guerrillas, and in World War II such American units as Darby's Rangers and Merrill's Marauders used many tactics of guerrillas.

The guerrilla isn't weaker than the larger force, just stronger in different ways. Guerrilla warfare is called "the war of the flea" because, like a dog and a flea, the dog has too much body to protect and the flea is too little to catch.

As Prais breaks the evolution of powerful firms into three stages, so we can trace the five stages of the evolution of the guerrilla "flea" corporation. The first finds the guerrilla beginning as a small local band of irregulars in a remote province—one local restaurant or a store, hotel, or a minor corporation. In the fifth the guerrilla no longer fights as a guerrilla but as a regular army or big company. Mao Tse-tung followed these stages and won China. Ray Kroc started McDonald's with one restaurant when he was fifty-two years old, followed the same stages, and created a commercial empire. Guerrilla Sam Walton opened his first store in Newport, Arkansas, and is now a billionaire, the richest man in the United States. His Wal-Mart strategy is simple and clear. Sell good brand-name merchandise in small-town America at low prices. Be nice to the customer, nice to your "associates," and keep costs down.

STAGE 1: BEGINNING IN A REMOTE PROVINCE

A guerrilla movement begins in a remote province because that type of area is the most inaccessible to the powerful armies, powerful corporations. That province becomes the guerrilla's base of operations.

In business warfare, too, guerrillas arise in remote areas, from small beginnings. For example, Frito-Lay Inc. and Pizza Hut were both started on loans from mothers to their sons of less than five hundred dollars. At the age of twenty Morris Siegel opened a health-food store, expanded into a cottage-industry operation producing one product, herbal tea, and was a millionaire by the age of twenty-six. His remote province was Boulder, Colorado. By the time he was thirty-four his company, Celestial Seasonings, had attained sales of twenty-seven million dollars. One year later Celestial Seasonings was sold to giant Dart & Kraft for an estimated forty million dollars.

The "war map" of the guerrilla begins with just a few dots where the guerrilla is strong and the larger competitor weak. Being more puny at this stage, the guerrilla avoids competing toe-to-toe with the larger force.

STAGE 2: WINNING THE LOYALTY OF THE PEOPLE

Powerful companies can exploit their three Ms—money, material, and manpower. The small guerrilla company hasn't a ghost of a chance of pitting itself against the large force's material resources. I visited a friend in one of the world's leading banks on two occasions in one year. Both times large sections of the building had been redecorated and given new furniture. "We've got so much money we don't know how to spend it," my friend said. "So we throw down some new carpeting." The guerrilla operates totally differently. He tosses quarters around as if they're manhole covers.

The only hope of winning the guerrilla has is to forget about beating the competition materially and to focus on the one thing the guerrilla needs to win. He accentuates intangibles, predominantly service, to gain the loyalty of his customers. Military history tells us that it's almost impossible to stamp out guerrillas in remote areas as long as they have the support of the people. That's why in every small city or town there's an independent store that's far and away more popular than the chains, restaurants that are busier than the franchises, a hotel that does better than the national names.

Debbie Fields started her cookie company in Palo Alto, California, in 1975 when she was twenty years old. To attract customers, she walked the streets, giving her cookies away. Shunning advertising but counting on word-of-mouth among loyal customers, she has since opened 315 stores to sell her Mrs. Fields chocolate-chip cookies.

The business guerrilla survives because of the quality of his customer relations just as the guerrilla in war continues to exist only because he has won the battle for the hearts of the people who support him. The people like the guerrilla. They're on his side; they give him food. When government troops come looking for him, they hide him. For the guerrilla the worst possibility is losing the backing of the people. He can't risk offending or mistreating the common man or woman. He tries to be fair and upright in his dealings with the people. If he takes something, he pays for it. Avis's famous motto "We're number two; we try harder" was a famous guerrilla tactic designed to shift our hearts to the side of the underdog. Chuckles Company, the nation's third-largest jelly candy maker, finds that some of its consumers feel loyalty to its candy out of "patriotic" sentiments because Chuckles were included in GI rations in World War II and Korea.

A purely military solution to guerrilla war is an impossibility. The only hope a government has of defeating a guerrilla is not by winning a battle or campaign but by winning the hearts and minds of the population away from the guerrilla. And the only way a corporate guerrilla can be beaten in the field is if the larger company snatches away the guerrilla's customers by winning their loyalty.

Will the sun rise tomorrow? It always has. Will a powerful company move in to attack the successful guerrilla? Same answer.

In war it's wise always to expect to be attacked sooner or later. You can expect it sooner rather than later if you're a small company that's making money. In this sense the appearance of a competitor is a direct function of your prosperity.

An athlete's "game face" is the look of tough, all-business determination. Guerrillas use the term "encirclement face" to describe the look of a frightened person. The larger force is forever trying to encircle and crush the guerrilla, and the guerrilla is endlessly struggling to avoid encirclement. Encirclement is the worst thing that can happen to any guerrilla, whether in warfare or business. The pattern of encirclement is this:

The guerrilla company shows itself to be prosperous. The larger company tries to stop it by offensive encirclement—keeping its influence localized, closing in, competing, taking away its business. Often the big company's strategy is to buy the loyalty of the people out from under the guerrilla by cutting prices. This is the "Magsaysay strategy."

When Ramón Magsaysay took over the Philippine antiguerrilla war in 1950, he paid large bounties for every weapon turned in and put rewards on the heads of key guerrilla leaders. His plan worked effectively, and the guerrilla movement never got out of the jungle.

In 1982 Vlasic Foods attempted to use the Magsaysay strategy against the guerrilla Farman Brothers Pickle Company in the Seattle area, Farman's base area. A subsidiary of the Campbell Soup Company, Vlasic has access to its owner's distribution, manufacturing, and advertising power. Vlasic's strategy is to rip into a market through price reductions. Against Farman, Vlasic dropped its price for a forty-six-ounce jar of pickles from $1.89 to $1.19, then to a rock-bottom seventy-nine cents, a price probably below cost of production. Vlasic also threw money into a heavy advertising campaign. Rather than lowering its price, Farman's stood fast, counting on its reputation for quality and the loyalty of area retailers to deal Vlasic a defeat. Farman's strategy worked; Vlasic's Magsaysay's didn't. Three years later Farman's was still controlling ten times the Seattle pickle-market share held by Vlasic. Vlasic's massive advertising had worked to Farman's advantage by inducing more people to buy pickles—Farman's.

But underpricing is not the only way to encircle the guerrilla company. Recycled Paper Products and Blue Mountain Arts make greeting cards. They're two of the estimated 500 guerrilla companies in the $3.6-billion industry. Hallmark is the giant, holding 50 percent of market share. When Hallmark saw the success of the two guerrillas in capturing the younger card-buyer market with offbeat, nontraditional, less straitlaced cards, Hallmark launched its encirclement campaign by matching their cards with its own imitations. The sales of the guerrillas fell off immediately. Losing ground, Recycled Paper and Blue Mountain filed suit against Hallmark. Hallmark calls it good business; the guerrillas call it copyright infringement.

In the first stage of its countercampaign against the opponent's "encirclement and suppression" campaign the guerrilla turns to the defensive and fights just to survive. Strategically you may be temporarily on the defensive against the larger force. But tactically you should launch short, sharp, hard-hitting campaigns, raids, and battles: short, sharp price cuts, for example, or hard-hitting advertising campaigns or sales blitzes. Guerrillas should always aim for the tactical offensive even in the strategic defensive.

In the second stage of the countercampaign the guerrilla takes the offensive and fights back. If the guerrilla is as successful as Farman's pickles was against Vlasic, the larger force takes the defensive and is forced to retreat. For example, the large competitor may attempt encirclement and suppression by cutting its prices again. The guerrilla company could swing to the counteroffensive by maintaining the price of its competing product and offering another brand below the attacker's price. The growing "flower consciousness" among Americans has increased the sales of exotic flowers here, but the small mom-and-pop retailer hasn't seen the benefits of it. Supermarket chains are selling flowers, offering comparable prices, but are open longer hours and stocking larger floral displays. Realizing that he couldn't compete on the supermarkets' terms, John Felly, a store owner in Madison, Wisconsin, experimented with plant nutrients to extend the life of his flowers. He'll give you your money back if the carnation you buy at his store doesn't last twenty days.

The larger forces' encirclement and suppression campaigns and the guerrilla's countercampaign against them—that's the "repeated pattern of the guerrilla's war." The pattern stops when the guerrilla becomes stronger than the opponent. Then the guerrilla shifts to the strategic offensive alone, encircling and suppressing the foe—and the opponent then resorts to countercampaigns.

The guerrilla must never attempt to confront the larger company except when and where victory is assured. Stay highly localized. Let the larger firm come to you. Defeat him through your strength— service to the people in one line of business or in one small area.

America's fourth-largest drugstore chain in total number of stores (1,263), Walgreen's is number one in sales and earnings. Its profits are more than twice that of Revco, which leads the industry with about 1,900 stores. Considered one of the nation's best-run companies, Walgreen's stays almost exclusively where it knows its chances of winning are best—in Chicago, in small-town Illinois, and in pockets of Indiana—places it knows well and where it is known.

Leonard Lavin walked into a Walgreen's store in 1940, saw a display of cosmetics, and decided to get into the health and beauty field. He called around and landed a job as a salesman. Fifteen years later he bought a small beauty-supply company that made more than a hundred products. He liked only one—VO5 Conditioning Hairdressing. That's the one he decided to push with heavy advertising. "The big boys in the industry had ignored us and thought we were another start-up that was going to fail, but we didn't," said Lavin. In 1985 Alberto-Culver celebrated its thirtieth birthday with annual sales of better than $386 million, ten times greater than its 1955 sales. In those three decades it evolved from a small local guerrilla company to a firm whose products are sold in more than a hundred countries.

STAGE 3: MOVING INTO NEW AREAS

The front is everywhere.

Guerrilla maxim

The guerrilla fighter, or guerrilla corporation, lives by mobility and fluidity. It's always moving away from where the competition is strong. It flows with the opportunities that arise in the marketplace.

The extreme version of the opportunistic guerrilla is the "fad marketer," the maker of the Pet Rock or the Rubik's Cube or the Wacky Wallwalker, for example. The fad marketer doesn't put much store in the ten- or twenty-year product plans of the giants but sees the game to be to excite the consumer temporarily—to move products with short life spans into the marketplace with dazzling speed.

Could the guerrilla fad marketer function in traditional corporate settings? Yes. He or she would see *every* product, *every* service, as needing fad elements to punch up sales periodically. For example, Colgate registered quick gains in the toothpaste business with its pump dispenser, a faddish innovation. Crest was forced to hurry to the market with its own pump.

Fad marketers are the extreme examples of the guerrilla's ability to flow with the opportunities appearing in the marketplace. But any company that shifts *toward* an opportunity and *away* from competition at the same time illustrates the essential quality of the corporate guerrilla. Efco Corporation is one example. Efco is a small guerrilla company of 420 employees making twenty-five million dollars in yearly sales manufacturing metal windows. Originally in the residential window business, Efco found it couldn't consistently make a profit. The price cutting was fierce, and there were many competitors. Efco merely shifted its emphasis, moving into the commercial window business, which requires a more sophisticated expertise but has far fewer competitors. Mao wrote, "We generally spend more time in moving than in fighting. . . . All our 'moving' is for the purpose of 'fighting,' all our strategy and tactics are built on 'fighting.' Nevertheless there are times when it is inadvisable for us to fight." It is usually inadvisable to fight when the force confronting you is too large or is strongly entrenched. If the competition is strong in a market, move away from it to another base of operations. If the competition doesn't come in to the base area, you have won it; it's yours.

Some guerrillas—probably most—are perfectly content staying in their base area and have no desire to venture out beyond it. The McDonald brothers from whom Ray Kroc bought the rights to franchise McDonald's restaurants were like that. They enjoyed sitting out on the porch of their home at night and watching the sunset. They didn't need any more problems, they said.

The guerrilla business person reaches a major choice point at the end of stage 3 when he or she asks, "How big do I want to be?" If the answer is "Very big" or "As big as I can become" or "I have no idea, but I'm going to go forward and find out what happens," the guerrilla will make a transition into stage 4. On the other hand, if the answer is "I'm content with what I've already achieved," the guerrilla will attempt to consolidate his or her position, protecting what's already been gained but refusing to expand much further.

As the guerrilla company grows stronger, it can begin to divide its forces and to advance into new geographic areas. Like Wal-Mart, it can open new stores, or, like Coors, it can spread farther east. Sharper Image moved on from its successful mail-order-catalog business (1985 sales over a hundred million dollars) into retailing. Now the guerrilla is establishing bases over a wide territory. It isn't expanding wildly but carefully. It moves into areas in which the competition isn't powerful. It stabilizes there and moves on. The war map of the guerrilla now has more dots on it.

STAGE 4: NIBBLING TO THE COMPETITOR'S PERIPHERY

The guerrilla expands its bases more, coming very close to the periphery of the larger companies' territory and putting pressure on them. Instead of a "no-man's land," we have a "both-man's land," a gray area visited by both the larger force and the guerrilla's irregulars. A kind of equilibrium sets in. The larger company now realizes that it can't eliminate the guerrilla but believes it can contain him. The guerrilla occupies himself nibbling away at the opponent's perimeter, taking some of its towns, its accounts, its market share.

Periodically the equilibrium is broken by skirmishes and battles. The big company wearies of the pesky flea and wants to put the cabash on him once and for all.

Guerrilla battles aren't prolonged. They're quick decisions. We know the book *The One Minute Manager*. Mao Tse-tung advocated "the five-minute attack" for guerrillas—a sudden assault, ferocious fighting, then an instantaneous break of contact before the opponent could exploit his strength—substantial resources, technology, and power—and bring the weight of his material and numbers to bear.

All corporations, really, are trying to fight wars of quick decision—to get to the marketplace first and to avoid costly, protracted warfare with competitors. "No conquest can be finished too soon," said Clausewitz. Sieges are always uneconomical. A small guerrilla firm, more so than its larger competitor, has to make certain it wins quickly. It can't afford not to.

A quick and favorable decision is not something you wish for but prepare for. Preparation is the first condition of a quick win. For example, because it can't afford to keep a big army in the field, Israel fights lightning-quick wars based on meticulous preplanning and preparation. On June 5, 1967, the Six-Day Arab-Israeli War broke out. Israel was attacked by the armies of Egypt, Jordan, and Syria. At the moment of the outbreak Israel had a standing army of only 50,000, but less than forty-eight hours later it had almost five times that number, 235,000 troops in the field, at the right places and ready to fight. By preparing beforehand, taxis, private cars, buses, and civilian trucks were organized to take reservists to preplanned rendezvous points where each person was assigned to a specific battle sector. Syria quit fighting after forty-eight hours of actual combat, Jordan after sixty hours, and Egypt after seventy-two hours.

Here are five guidelines for fighting "five-minute" guerrilla attacks.

- Seize the opportune moment.

- Concentrate superior forces. A guerrilla will suffer every time it fails to concentrate its forces. If you're less than half as strong as the enemy, says Frederick the Great, "act only with your entire army. Do not detach any unit." Concentration is the heart and soul of guerrillaism. Concentrate your forces in a segment, a terrain or area of expertise so that you are the leader —the leading chain of grocery stores in Pleasantville; the best accounting services or the major construction company or trucker.

- Outflank the competition.

- Operate on favorable terrain. (Fight on ground of your choosing and from positions where a relatively few determined people can stall an army.)

- Attack the competition when it hasn't yet consolidated its position.

Unless these five guidelines are followed, you won't be able to achieve a quick win in a battle or campaign.

STAGE 5: LAUNCHING THE GENERAL OFFENSIVE

This critical phase begins when the guerrilla and the larger company have reached a balance. Now the guerrilla completely changes his strategy and tactics. He seizes the initiative not as a number of bands of irregulars but as a *regular army* fighting conventional warfare. In the past, as a guerrilla, he had avoided battles and used hit-and-run tactics. Now he does battle. He uses regular troops. He follows the lessons of conventional war, particularly the lesson of concentration. He boldly throws superior numbers into concerted attacks against the other companies' weakest points.

The war map of the guerrilla began with a few isolated dots, then added dots (new territory). Now entire large areas belong to the guerrilla. Wal-Mart's Sam Walton started with just one store. Now he has ninety thousand people working for him.

TWO ERRORS TO AVOID

If you're a guerrilla company there are two fatal errors to avoid:

1. The first error is losing contact with the population—the consumer in your terrain. The moment the guerrilla company starts neglecting those little extras that brought him business in the first place, he's in trouble. In a room of a Helmsley Hotel you'll find a note suggesting that if you've forgotten to pack a hair

dryer, you call the desk and borrow one of theirs. And you'll find complimentary shower caps, bottles of skin moisturizer and shampoo, candy, a clothesline over the tub for drying your swimsuit, etc. That's guerrilla service. If Helmsley ever starts skimping on those nice little extras that are all the nicer at the weary end of a long day of traveling, that small but growing chain will be just another hotel.

The government of France owns Renault, the car maker; Renault, in turn, owns 46 percent of American Motors Corporation and runs it. Renault has two population problems. It bought a large share of AMC hoping to gain a foothold in the American auto industry but never achieved success here. One AMC dealer complained of having to pay consumers fifty dollars as an inducement to test-drive a car. Between 1978, when Renault invested heavily in AMC, and 1986, AMC lost almost three-quarters of a billion dollars. Renault also had serious population difficulties in Europe. At one time the leading European automaker, Renault lost 25 percent of its market share there between 1980 and 1986 and three and a quarter billion dollars in 1984–85. The French government wishes to sell it to a private corporation, but even the prime minister has asked who would want to buy it now?

2. The second fatal error you must avoid if you're the guerrilla is attempting to shift from guerrilla warfare to conventional warfare too quickly, before you're strong enough to battle directly with the competition. Dreyer's Grand Ice Cream has positioned itself in the market niche above lower-priced supermarket brands and below superpremiums like Häagen-Dazs and Frusen Glädjé. While Dreyer's, now sold in thirteen states west of the Mississippi, is moving into the Midwest, its owners are not opening their own ice-cream parlors but selling their product through restaurants and stores, thus taking the course of least resistance against "dipping parlors" like Baskin-Robbins.

On the other hand, Cook's Mart Limited, the gourmet cookware chain, crumbled into liquidation after eight years when it made the devastating guerrilla error of attempting to expand too fast.

Any guerrilla that commits these errors—whether corporate or political—will lose.

GUERRILLA-STYLE MANAGEMENT

*The advantages are nearly all on the side of the guerrilla
in that he is bound by no rules. . . .*

Sir A. P. Wavell

Recently I found myself in the birdhouse of a zoo looking at a glassed-in cage of lovebirds. Where you find one lovebird, you find another, its mate. The two perch closely together. If one moves, the other does, too. They're always paired up and intimately connected. In the same way, guerrilla strategy and tactics are invariably accompanied by a particular system of management. Whether you're in charge of an entire corporation or a division or one department, if you're competing guerrilla-fashion, you'll want to use guerrilla-style management.

- *Guerrillas emphasize decentralization over centralization.* Dynamism, initiative, speed, maneuverability, and quick responses to new situations are the qualities that gain the guerrilla victory. Slow the guerrilla irregular down with a "check with headquarters first" mentality and you will lose guerrilla competitiveness. Even some of the nation's largest companies exploit the advantages of decentralized, self-contained, guerrilla-style units. Panasonic and Quasar decentralize most functions to create small, semiautonomous, market-sensitive enterprises. Esmark and Northwest Industries manage their corporate portfolios similarly—unit managers operating autonomously while a lean corporate staff allocates money and evaluates the business mix. Hewlett-Packard and 3M, although huge companies, emphasize a small-business decentralization. When a division becomes too large, they "clone" it, splitting it into a smaller size. Sears centralizes merchandising but decentralizes store management.

Napoleon eventually wanted to make *all* the decisions, and so did Henry Ford. Ford went so far as to have spies report on managers who dared to make independent decisions. Neither Ford nor Napoleon could function in a guerrilla-style company. You can't make decisions for the guerrilla from headquarters.

- *Guerrillas are more oriented to the offensive than any other kind of company.* Guerrillas are faster on their feet and quicker to perceive and exploit trends and opportunities. Even the largest companies appreciate this special skill of the guerrilla-entrepreneur to sense an opportunity. For example, John Akers, CEO of IBM, talks about the "new IBM." He means a style of operation that runs counter to that of firms that dominate an industry, including long time gaps between new products. The new IBM will operate, says Akers, like a company of entrepreneurs continually incubating new ideas. IBM has formed independent business units, separate companies designed to explore new markets. Some of them will fail, says Akers. "In fact, you ought to fail in some of them." IBM's reputation for being big yet fast moving and flexible at the same time has earned it the nickname "the elephant that dances." 3M managers are held responsible for bottom-line management and for introducing a number of new products every five years. PepsiCo has vowed to become the largest company in the world with a small-company entrepreneurial mentality. And Motorola chairman Robert Galvin has observed that the company's greatest product successes are as likely to come from twenty-five-thousand-dollar research projects as from hundred-million-dollar projects.

Many of the themes that major corporations are just now learning the guerrilla has long understood, including:

- Small is better than large.
- Narrow market segments are more profitable than broad ones.
- Originality and innovation are better than business as usual.
- Temporary is better than permanent.
- Short chains of command are more productive than long.

- *Guerrillas value each of their fighters individually.* In Israel, for example, the prime minister is immediately given the name of each casualty. Although it has grown far beyond its original guerrilla status, Wal-Mart continues to treat its employees like irregulars who are all-important. Wal-Mart's hourly workers are regularly visited by company executives, including CEO Sam Walton, and asked, "How's your manager treating you? Do you have any ideas?" Says one employee, "It makes you feel like a million dollars." Wal-Mart regional managers spend four days every week visiting their stores. On Friday and Saturday they're back at headquarters to report to top management. After Saturday's meeting they telephone their district offices.

  Limited Incorporated, the clothing retailer, owns twenty-four hundred stores. It maintains a zeal for entrepreneurialship by giving its buyers and others the authority to make decisions fast, without having to check "upstairs" by emphasizing employee ownership of the company, and building in-group loyalty among employees.

  Du Pont believes "every employee must be a salesman." Almost every corporation in the country says that; Du Pont means it. Recently the company asked for volunteers for its "Antron army" to market its consumer-credit plan to stores selling Du-Pont's Antron-fiber carpets. Five hundred stenographers, clerks, and factory hands volunteered, only 10 percent of whom had ever sold anything before.

  Fifteen years ago Quad/Graphics Printing Company of

Pewaukee, Wisconsin, had ten employees, one press, and printed two hundred thousand copies of one magazine, *Fishing Facts*, a month. Now it prints sixty million magazines each month, ranging from *MAD, Playboy, Ms.,* and *Mother Jones* to *Time, Newsweek,* and *U.S. News & World Report.* In one recent year it developed one new contract every eleven days. One book recently named it one of the hundred best places in America to work. Says Harry Quadracci, Quad/Graphics founder, "We have achieved extraordinary results from ordinary people." Like a guerrilla fighter, Quadracci has no organizational charts, no job titles, no job descriptions, no time clocks, and doesn't believe in writing policies and procedures down on paper. Quad/ Graphics has increased sales by at least 50 percent in every one of its fifteen years of existence.

- *Guerrillas fight their flea wars for a long time.* All guerrillas who win do so by wearing the competition out. A guerrilla wins gradually. With a machete he captures a rifle. With fifteen rifles he captures a machine gun. With fifteen rifles and a machine gun he takes a convoy carrying seven machine guns and eighty thousand rounds of ammunition. Guerrilla managers don't expect to become giants overnight.

- *Guerrillas encourage "positive indiscipline."* Positive indiscipline is employees' bending the rules not because they're poor workers but because they're very good. It's indiscipline from excessive zeal and drive. Old-line corporations often try to discourage it; guerrillas actively encourage it. Guerrillaism is always anti-bureaucratic.

  The great victory of the Germans over the Russians at the Battle of Tannenberg in 1914, a double envelopment, was credited to General von Hindenburg and Chief of Staff Ludendorff but was really the result of the "positive indiscipline" of a young operations staff officer, Col. Max von Hoffman, and a decisive corps commander, Hermann von François. On numerous occasions during the battle, von François completely disregarded Ludendorff's vacillating orders. Ordered by Ludendorff to end his flanking move, von François lied, claiming his artillery was delayed, and refused to obey the absurd order.

Dependent staff members don't make waves, and independent thinkers may, but in the heat of battle the independent thinkers often are the people who bring you big wins.

Achieving Security

SIGNS

*If I am able to determine the enemy's dispositions while
at the same time I conceal my own, then I can
concentrate and he must divide.*

Sun Tzu

In his book *The Art of War*, Sun Tzu claimed that "what enables
the wise sovereign and the good general to strike and conquer, and
achieve things beyond the reach of ordinary men, is foreknowl-
edge. . . . Therefore, determine the enemy's plans and you will know
which strategy will be successful and which will not." He adds that
if you know yourself and you know the enemy, you will *always* win.
On the other hand, if you know neither yourself nor your enemy,
you'll *never* win. To Sun Tzu a great part of the art of war consists of
perceiving and correctly interpreting SIGNS, including these:

- If you hear the enemy firing weapons in his camp, it's because
  he's discharging and cleaning his weapons. Expect an attack
  tomorrow.

- When the enemy stays back and hurls insults, he's trying to provoke an attack. Be careful of being lured into a trap.

- Humble words coupled with increased preparations are signs that the enemy is about to attack. On the other hand, violent language and driving forward as if he's going to attack are signs that he will retreat. (American manufacturers have noticed that Japanese companies who are normally hospitable to conducting tours of their facilities to visitors develop closed-door policies shortly before they introduce a new technology or product.)

- Whenever there is suddenly a great deal of movement in the enemy camp, it means the critical moment of attack is coming.

- "If those who are sent to draw water begin by drinking themselves, the army is suffering from thirst." (If the competition's staff starts looking for other jobs, it's a sure early sign that he's in trouble.)

- "If the enemy sees an advantage to be gained and makes no effort to secure it, the soldiers are exhausted." (If you see where the competition could gain by doing X, you can assume that the competition has seen it, too. If he doesn't take advantage of X, he's undergoing some problems.)

When you're thinking of going to war, study the signs, says Sun Tzu. If you don't, you could be in hot water. Chinese intervention in the Korean War began when the Chinese moved three hundred thousand troops into North Korea without the U.S. Air Force noticing. And when interpreting the signs, only rely on people whose judgment you trust. Union general George McClellan had a true genius for making sure his troops were well trained. His only weakness was that he was afraid to fight. One reason for his tendency to do nothing was the exaggerated reports of Confederate strength supplied to him by his intelligence agents—employees of the Pinkerton Detective Agency, who were completely ignorant on matters of war.

More than two thousand years separate Sun Tzu from Harold Geneen, the legendary former CEO of ITT. But Geneen came to the same conclusion about the importance of *knowing*. "Facts are power," he wrote. However, he adds, what are called facts are often not facts at all. In his book *Managing*, Geneen describes how early in his tenure with ITT he discovered that there were "apparent facts," "assumed facts," and "reported facts," each of which was accepted as fact by managers but in most cases were really not facts. The highest act of management, he adds, is the ability to smell a real fact, an "unshakable fact," from all the phony facts.

Clausewitz was as skeptical about facts as Geneen. He believed that three-quarters of the things on which decisions were based in war were obscured by the fog of war and that most of the information available to the commander was contradictory, even more was untrue, and more still was questionable. In business warfare, every decision you make implies a forecast. The fog of war doesn't blanket all forecasts but certainly many of them. Some things you know will happen. They have before; they will again. Others you know will occur sometime, someplace, in some form, but you're not exactly certain of when, where, and how. Other situations you feel are going to happen, but you're not certain they will. Finally, shrouded deep in the fog of business warfare are those things you can never forecast until they happen. They're completely unpredictable and can't be foreseen. In Clausewitz's experience the fog of war could never be completely eliminated, just reduced through good intelligence systems and skillful human judgment—the type of judgment to smell a real, unshakable fact that Geneen claims is the "highest act" of management.

Thirty years ago Arthur Nielsen, Jr., then president of A. C. Nielsen, listed the thirteen most common marketing failures. Almost all of them result from problems with what Sun Tzu would call correctly perceiving and interpreting signs:

1. Failure to update your product

2. Failure to estimate the product's market potential accurately

3. Failure to perceive the market trend

4. Failure to consider regional differences

5. Failure to appreciate seasonal differences in buyer's demand

6. Failure to establish the budget for advertising on the basis of the job to be done

7. Failure to stick to long-range goals

8. Failure to test market fresh ideas

9. Failure to distinguish between short-term tactics and longer-range strategy

10. Failure to admit defeat and to change

11. Failure to attempt new approaches while a brand's sales are climbing

12. Failure to coordinate and integrate all elements of the marketing push into the overall program

13. Failure to estimate objectively your competitor's brands

The principle of war covering Geneen's cry for the "unshakable facts," Clausewitz's search for clarity even in the fog of war, and Sun Tzu's "signs" is the PRINCIPLE OF SECURITY. The main objective of security is to protect your freedom of movement, "to act securely and with certainty, whatever the enemy may do."

Since the unknown is the governing condition of war, the first element of security consists of pressing as far into the fog of war as possible by securing the greatest amount of intelligence and piecing it together to form an accurate picture. According to Napoleon, the objective of intelligence is "to foresee everything the enemy may do, and be prepared with the necessary means to counteract it." To French World War I general Ferdinand Foch, security involved knowing enough about the adversary that you could avoid his blows, prevent him from surprising you, and keep him in the dark about your plans as long as possible. This competition within the competition of business warfare is primarily for intelligence—accurate and useful information on your competitors, your competitor's leaders, and consumers.

KNOW YOUR COMPETITORS

The unknown is the governing condition of war.
Ferdinand Foch

Englishman B. H. Liddell Hart was wounded at the Somme in 1916. After a period of hospitalization he returned to the army for light duty, revising the English infantry training manuals. Most people who have written a training manual can hardly imagine a more boring thing to do except perhaps writing a proposal. Liddell Hart was different. The job so engrossed him that it set him on a course he would follow the rest of his life. Long after he had retired as a captain, he continued his exhaustive studies of strategy and tactics and eventually became recognized as the twentieth century's most famous military historian and philosopher of warfare whose views of war have been studied and applied by many of the great commanders of our time. When he died in 1970, Liddell Hart was known as "the captain who teaches generals."

Liddell Hart's search for the maximum simplicity brought him to the idea that all of warfare from tactics to strategy could be likened to a single unarmed "man in the dark" encountering a single unarmed adversary. To win, one would have to discover the other, reconnoiter, attack decisively, and exploit any advantage.

The man in the dark reaches out with one arm and gropes for the enemy (discover). Touching him, he feels around till he finds the other's throat (reconnoiter). With his free hand he delivers a knockout punch (decisive attack). Before the enemy can recover, the man in the dark follows through, rendering his enemy powerless (exploit).

Corporations are like the unarmed "man in the dark" as they confront their competitors, and corporate managers are responsible for *discovering* the competition, *reconnoitering* them, *attacking*, and *exploiting* competitive advantages. Chapter 5 of *Waging Business Warfare* dealt with attacking and exploiting advantages—the military principle of offensive action and the subprinciple of pursuit. The principle of security involves discovering and reconnoitering the competition. The Duke of Wellington claimed to have spent his entire career guessing what he might meet beyond the next hill or around the next corner. Businessmen and women do the same. They all spend part of their time wondering what the future holds, what the market will do, what competitors are plotting, and how they will respond to strategic moves on your part.

You're trying to learn the competition's *strength, weaknesses, tendencies*, and *intentions* as best you can. At the same time, you're trying to keep him ignorant of your strengths, weaknesses, and intentions as well as you're able to. "Success in war," wrote Francesco Guicciacardini, "is obtained by anticipating the plans of the enemy, and by diverting his attention from your own designs."

Analyzing Your Competitors

Anyone can find numerous cases of industries in which one or more of the leading companies are obviously stumbling because they do not have a clear picture of the competing strategies they face, or what their own capabilities are to counter competing strategies.

Charles R. Wasson, *The Strategy of Marketing Research*

Analyzing your competitors needn't be highfalutin; it can begin by simply asking around. John Grubb is the owner of the San Francisco-based construction company Clearwood Building, Inc. Together with his brother Robert, he systematically asked the firm's architect-customers what they felt were the shortcomings of Clearwood's competitors. They were told poor manners, dilapidated construction trucks that upper-class customers didn't want parked in their driveways, and thoughtless workers who tracked dirt into houses. The Grubb brothers used this competitor intelligence to mold their company into *the* contractor for the Bay Area's more affluent customers. The Grubbs purchased a new truck and kept it immaculate; their job estimators wore jacket and tie, and their workers, trained to be polite, laid protective runners over carpeting before entering a prospect's home. Within two years the company's annual income leaped 500 percent, from two hundred thousand dollars to one million.

To analyze your competitors, you have to discover who they are. This isn't always so obvious as it appears.

- *Potential competitors* aren't on the horizon, but hidden on the other side of it. You can't see them at the moment, but poor pricing or bad customer service will bring them pouring down on you.

- *Indirect competitors* offer products or services that the consumer can substitute for yours. At this moment the competition of one airline—American, for example—is other airlines. But when telephone companies succeed in reducing the expense of teleconferencing to affordability, all the airlines will have a new competitor to deal with.

- *Direct competitors* are usually obvious. There's no secret; you know who yours are.

Harvard University's Michael Porter describes a "framework for competitor analysis" that consists of analyzing four main components —goals, assumptions, strategy, and capabilities.

1. Diagnosing competitors' GOALS. Being aware of goals enables you to gauge how content each competitor is with its current position in the industry. Also, knowing what their goals are can give you some clues as to how threatened competitors may feel by competitive moves on your part.

2. Identifying competitors' ASSUMPTIONS—about themselves and their situation, the industry, and other firms in the industry. The saying goes, "Oh, to see ourselves as others see us." This component of competition analysis is slightly different. Its purpose is "to see the competitor as it sees itself." One night I went to dinner with the marketing director of one of the world's largest computer-software companies. As we ate our appetizers, he did a great deal of talking, mentioning among many other things the need for any computer-software company that wanted to grow to advertise. During the main course he said as an aside, "We're in a growth mode." When dessert

came, I said, "So tell me about the new advertising program you guys are coming out with in the near future." All I had done was to listen to his assumptions.

3. Understanding the competitors' CURRENT STRATEGY.

4. Estimating each competitor's CAPABILITIES—its competitive strengths and weaknesses in products, distribution, marketing, research, costs, finances, production facilities, management agility, organizational design, etc.

Here are some of the specific questions you'll ask yourself about your competition.

- How does the company view itself? The industry?

- Who is likely to respond to your move? What form might their responses take?

- What are their present objectives and strategies for reaching them? What are they after? What might be their future objectives?

- Are they risk takers or not?

- What results are they achieving in sales growth, rate of return, etc.?

- If they're part of a larger company, what is the strategic emphasis of that company?

- What are the company's strengths and special capabilities?

- What are its weaknesses and vulnerabilities?

- How does it concentrate and specialize its forces—by terrain, strength, or competitor's vulnerability?

- What is their core way of doing business, their dominating strategy, and how is this reflected in technology, price-product strategy, promotional strategy, and every other aspect of their doing business?

- Which market segments are they successful in entering? In which do they do particularly well; where do they not fare as well?

- Who are the potential competitors on the horizon?

A useful competitor analysis results in a profile containing the answers to these questions and any other information you can gather. You don't have to be a giant in your industry to benefit from such profiles. Dallas's Miller Business Systems, Inc. enters competitor information into computerized profiles, then analyzes the profiles for possible competitive advantages. While browsing over one competitor profile, Miller's vice-president of sales and marketing noticed that the company had put on nine furniture salesmen in just ten days. Correctly seeing this as a sign that the company was making a move into the office-furniture market, Miller had its sales force concentrate on making additional sales calls and successfully defended against the competitor's offensive.

Create an "Enemy Department"

Though thine enemy seems like a mouse, watch him
like a lion.

Italian proverb

Most every marketing plan has a section on competitors, but it's rarely very detailed or particularly useful. And even the smallest company keeps a manila file folder on "the competition," but it's usually just a dust gatherer.

The basic quality that strategic thinkers all have in common is the ability to put themselves in the opponent's shoes—to ferret out the opponent's perceptions of the battle and their underlying assumptions, to glean their way of waging war or doing business, and then to use this knowledge to do what the competition doesn't expect or what they probably won't counter with force and determination. An enemy department (it could be one employee in your company or a team of executives who do it part-time) would be given the job of "becoming" the competition, analyzing your strengths and weaknesses, preparing competitive moves against you, a list of probable reactions to your offensive moves and devising other ways of outcompeting you.

For some years General Electric has maintained its version of an enemy department. Called the Business Environmental Section and located at GE headquarters, it prepares reports on probable environmental changes for use by corporate divisions. Other corporations have made "devil's advocates" a part of their planning processes. The devil's advocates are paid not to take the employing company's point of view but the competition's. In *The Prince*, Machiavelli suggests that even in peacetime, when out for a ride with his advisers, the Prince should ask them what they would do if the enemy attacked over that hill or from that direction. Some companies do essentially the same by conducting "aggressor" exercises to simulate what products and markets competitors might attack. Others, like Xerox, pay experts and consultants to prepare studies that may challenge its executives to think creatively about opportunities and threats. A number of companies have assigned managers the role of competitor "shadow." They keep a close eye on a particular competitor. During strategy sessions the shadow is to be able to describe how his or her company will react to strategic actions under consideration. ITT, under Harold Geneen, assigned a dozen to sixteen product-line managers to monitor the entire product lines of competitors of ITT subsidiaries. Probably every manager and certainly every management team should always be asking in a very Machiavelli-like fashion, "If I were that competitor, what would I be hatching? And based on what I know about us, what would I attack?"

In the military, "intelligence templates" are used to put intelligence information to critical use. They relate enemy activities to terrain and weather on the battlefield. They can be graphic drawings that can be placed over a map to show where the enemy is, how strong he is at every point, how widely spaced units are, composition of units, etc. Corporations that are concerned enough with competitors could develop their own "intelligence templates." They would take the form of written rundowns of competitions and contain answers to the questions raised earlier in this chapter, in the section "Analyzing Your Competitors."

Look for the Strength; Discover Their Weakness

All companies and all individuals have at least one weakness that you can take advantage of. Ironically, it's often to be found very close to the company or person's main strength.

Belisarius was a master of converting an enemy's strength into weakness. For example, it was obvious to him that the main strength of the Goths was power and superior numbers. They knew it, too, and made the direct frontal assault their *modus operandi*. Belisarius defeated them by leading them into a headlong direct assault that Belisarius, with a much smaller force, found little difficulty flanking and defeating. On the other hand, the Persians were subtle, careful thinkers. That was their strength. Understanding that any opponent who thinks too much tends to be wary of action, Belisarius took advantage of the Persian wariness through rapid bang-bang-bang maneuvers and won, again turning the opponent's main strength into a weakness.

Generally, a small firm's strength is speed of translating an opportunity into a business venture; and a large company's main strength is the ability to spend money without the need for immediate return. The chief weakness of both is near those strengths. The small firm can easily dissipate its chance of concentration and exploitation of any one opportunity by moving on to another too quickly, and large firms often become dollar arrogant. Believing that to win means to outspend the competition, they often throw good money after bad.

Read Their Mail

Someone once said, "Gentlemen don't read each other's mail." In *The Craft of Intelligence*, former CIA director Allen Dulles wrote, "When the fate of a nation and the lives of its soldiers are at stake, gentlemen do read each other's mail—if they can get their hands on it." Adman David Ogilvy came close. He once won a large account by reading the potential client's agenda upside down from across the desk and being able to answer each point before it was raised. There's a great deal of figurative mail reading in business warfare.

The Cannes Film Festival is not merely large-busted starlets and fun in the sun. It's big business, and the stakes are astronomically high. If a film distributor can find out the amount of competitors' bids on a film, he can simply tack on another few thousand dollars and walk away with potentially fifty million dollars in rentals. One company executive inadvertently leaked his firm's bid to a competitor by absentmindedly tossing a piece of scratch paper in a hotel wastebasket.

British Airways called in private investigators to uncover the executive who was leaking plans for routes and schedule changes to newspapers well in advance of planned release dates, and a Swiss chocolate manufacturer caught two of its employees peddling the company's recipes to Soviet, East German, and Chinese embassies in Berne.

General Electric, Digital, Ford, and Union Carbide are just a few firms that have developed sophisticated competitive intelligence systems. Kraft and Gillette have employees collect information on competitors and their products, and so do Rockwell International, J. C. Penney, Ford, and Del Monte, among others.

Salespeople get around. They're often used to pick up what scuttlebutt they can on competitors, trends, and consumer preferences. The use of commercial travelers to gather intelligence has a long tradition in warfare. In the thirteenth century, the Mongol Empire entered into a secret treaty with Venice. The Mongols would destroy every non-Venetian trading station wherever it was found, leaving Venice with a trading monopoly. In return, all traveling Venetian merchants would prepare and submit to the Mongols highly detailed reports of troop movements and economic strengths in all countries they visited.

Sound competitive information can be gained from the everyday grapevine and simple, Sun Tzu-like observations. What are they pushing in advertising; what are they selling and to whom? What are they changing; what's being kept the same? You can gain important information about competitors very inexpensively if you know where to look. You can secure product information by buying competitors' products and analyzing them component by component. The Freedom of Information Act makes available to you records, letters, test results, and other information submitted to governmental regulatory bodies. Personal contacts with financial analysts can also provide you with insights into competitors. You can hire on retainer a marketing research firm to provide you with current information on competitors.

Signs of possible competitors' moves can also be discovered by way of:

- Steady contact with information sources in the field. Suppliers and customers, for example.

- Attending conferences and shows.

- Monitoring newspaper and magazine articles—the *Wall Street Journal*, *Fortune*, *Business Week*, and your local newspaper, for example. News releases or "planted stories" are often clear signals of a competitor's intention—to market in a new territory, to buy a plant, acquire a company, etc. At times they're intended to scare you off from competing or to test your reaction, to express dissatisfaction with one of your moves, etc. Ordinarily, however, newspaper articles or items in a local business publication can be quite useful if your mind is fertile and you can read between the lines. What does X company's removal of two top executives mean? Why were they axed now? What could the action portend? etc.

- Analyzing speeches and writings of competitors' CEOs. Irvine O. Hockaday, executive vice-president of Hallmark Cards, went to see John R. Purcell, chairman of SFN, a textbook publisher and broadcasting company, to discuss Hallmark's purchasing the company. Hockaday "had studied us," said Purcell. "He had copies of every speech I had made to securities analysts, copies of articles. . . . That disarmed me."

- Financial statements and reports. These can often tell you if the competition is investing in your industry or diversifying out of it. If you simply buy one share of a competitor's stock, you will receive all of its stockholder reports.

- Agents. You can always question people who were once with the competitor but who have left them. Even the best places to work lose managers. On the average, managers leave large corporations at a rate of about 6 percent every year.

- Industrial tours. These provide the opportunity to secure information on competitors and potential competitors. Kellogg Company, the biggest cereal producer in the world, conducted free tours in its Battle Creek, Michigan, plant for eighty years, until April 1986. Kellogg put an end to them to protect itself against competitors' spies stealing production secrets. Kellogg discovered that competitors were sending their representatives on the tour repeatedly, then setting up competitive plants in foreign countries.

KNOW YOUR COMPETITOR'S LEADERS

It is essential to know the character of the enemy and of their principal officers—whether they be rash or cautious, enterprising or timid, whether they fight from careful calculation or chance.

Vegetius (A.D. 390)

Said an executive of Boise Cascade:

Our strategy sessions are very thorough. . . . Our strategies involve not only an analysis of our own actions within our present environments, but also an analysis of the reactions of others to our own actions. . . . We weigh all the things he can do. We try to decide what the chances are that he will do any one of them —recognizing that each may not be equally available. We consider his position, his balance sheet, his psychology, his gutty or conservative nature as an individual, and the past patterns of the corporation's behavior.

The term "psychological warfare" is redundant. *All* of warfare is psychological, including the commercial type—a battle of minds and spirits. Throughout history, all generals and corporate leaders of stature have come to the same conclusion: one of the main targets of the conflict is the mind of the other leader. Liddell Hart said that the most profound truth of all truths of war "is that the issue of battles is usually decided in the minds of opposing commanders, not in the bodies of their men." Bruce Henderson, noted corporate strategist, states that "victory, if achieved, is more often won in the mind of a competitor than in the economic arena." He adds that through its strategy, corporate management must convince each competitor to stop short of maximum attempts to gain customers and profits, and that convincing them to do so depends on emotional and intuitive factors rather than on reason.

The Duke of Marlborough should have lost the Battle of Ramillies (1706), and Marshal Villeroi, the commander of the bigger and stronger French army, should have won. The reverse happened. Villeroi advanced quickly on Marlborough, as the French king had ordered him to. But as he came close, Villeroi lost his courage, stopped, and swung to the defensive. Why? The *only* reason was Marlborough's reputation. Villeroi was intimidated into the defeat, which soon followed his loss of nerve. As Henderson notes in his passage, the cause of stopping short was "emotional and intuitive."

All great military commanders, as distinct from the merely competent ones, have had the ability to "penetrate his opponent's brain" and to vary their methods of war according to the individual commanding the opposing force, fighting one kind of war against one, another against a second. For example, Confederate general Robert E. Lee, a true leader, took far greater chances against Union commanders McClellan, Pope, and Halleck than he did against Ulysses S. Grant.

Hannibal's astonishing conquests can be attributed directly to his constantly molding his strategy and tactics against the character of the commander opposing him. He said, "It is to be ignorant and blind

in the science of commanding armies to think that a general has anything more important to do than to apply himself to learning the
inclinations and the character of his adversary." Hannibal took his
own words to heart, as we saw in chapter 6, when he used the single
insight that his adversary, Flaminius, was easily insulted to goad the
proud Roman into an ill-advised and fatal attack. Such dramatic results of the right—or wrong—reading of an opponent are also found
in business warfare. At certain times the future of an entire major
corporation, even of an industry, has hinged on the ability or inability
of a firm's leader to correctly or incorrectly judge the personality and
reactions of a foe.

On Friday, February 21, 1986, Texas Air Corporation's president,
Frank Lorenzo, submitted an offer to buy Eastern Airlines. Since
Lorenzo had the reputation for being adamantly antiunion, Eastern's
executives, particularly its president, Frank Borman, apparently believed that the Lorenzo bid could be used as leverage to force Eastern's
three unions to accept the airline's demands for 20 percent wage cuts.
The unions had only three choices: to accept the pay cuts, force Eastern into bankruptcy, or be taken over by Lorenzo. Borman's bet was
that labor would want to avoid the second and third choices and would
go along with the cuts. When two of the unions accepted the wage
concessions, it appeared that Borman had wagered correctly, but this
soon changed. Charles Bryan, head of Eastern's machinists union,
refused to agree to the cuts unless Borman resigned. Bryan refused to
budge from his position. Borman had miscalculated, and at thirteen
minutes to three on the morning of February 24, Eastern's board
agreed in principle to the Texas Air buyout for six hundred million
dollars. Less than seven months later—on September 15, 1986—
Houston-based Texas Air agreed to buy troubled People Express, Inc.
and its Frontier Airlines, a deal that, when completed, would put
Texas Air in control of the largest system of air carriers in the United
States and garner the airline 17.3 percent of market share compared
with United's 16.3 percent and American's 13.6 percent.

Competitions between armies and between one corporation and

another clearly have important psychological and emotional elements. Hardly anyone believes any longer that issues are really decided on the basis of totally rational decision-making processes alone. After conducting a comprehensive study, researcher Yehezkel Dror concluded that when they make decisions, people *almost never* spontaneously apply the elements of "rational" decision making such as searching for the best alternatives or considering probabilities. Instead of "rational man," we have sometimes rational but often irrational man. Business decisions are not made by machines but by real people. C. Roland Christensen, Kenneth R. Andrews, and James L. Bower of Harvard University's business school comment on the primacy of personal values in strategic decision making:

> We must acknowledge at this point that there is no way to divorce the decision determining the most sensible economic strategy for a company from the personal values of those who make the choice. Executives in charge of company destinies do not look exclusively at what a company might do and can do. In apparent disregard of the second of these considerations, they sometimes seem heavily influenced by what they personally *want* to do.

A corporation's strategy is a reflection of the personality of the company's leaders. "When you hire the man, you hire his strategy," said one investment banker. "And if you fire the man who created the strategy you fire the strategy." That's precisely what happened at Beatrice Companies in 1985. When Beatrice's board fired James L. Dutt as chairman in late summer 1985, it also lost his strategy for making the company the world's largest seller of brand-name products. Knowing who is in charge over there, what his or her track record is, and what he or she is like personally usually provides immediate insights into the competitor's probable strategic course of action. This is demonstrated time and again.

That Germany of World War II would follow a strategy of *blitzkrieg* was perfectly predictable given Hitler's unalterable belief in his personal destiny, which enabled him to take extraordinary risks.

Montgomery was always action oriented, and mobile war suited him. Patton was, too, and thus mobility was also his essential approach to war.

One of the most historic events in business warfare was the campaign started in 1968 by then minute MCI Communications Corporation against giant AT&T, eventually resulting in the breakup of Ma Bell. The man responsible was MCI's chairman, William G. McGowan. Aggressive war suits the man; it should not be considered astonishing that in the five years between 1980 and 1985, McGowan led MCI to an increase in sales of 1,200 percent. Such aggressiveness is contagious. Bert C. Roberts, MCI's president and CEO, says about himself, "I'm a doer. I make up my mind to do something. Then the goal is to do it and to get on to the next challenge." It's not coincidental that although AT&T is still far and away the biggest competitor in the long-distance-telephone industry, its market share is declining while MCI is clearly the fastest growing.

Joseph F. Alibrandi is the head of Whittaker Corporation, headquartered in Los Angeles. Already a conglomerate in chemicals, metals, and electronics, Whittaker is now spending $100 million to build a national system of HMOs. Why HMOs? Given the fact that Alibrandi is a distance runner and good health enthusiast, it's not completely surprising that he would steer his company into the health field. Alibrandi is also a patient man, supportive and forgiving, who believes that managers can't be expected to be right all the time. This same quality of patience is implicit in Whittaker's HMO strategy. He *expects* the venture to lose money for up to five years, and he's willing to wait for the profits.

More than fifteen hundred years ago Li Ch'uan, commentator on Sun Tzu's *Art of War*, identified three basic types of leaders and three fundamental forms their leadership would take. If the other side has a (type 1) valiant leader, expect a fight; if their leader is (type 2) cautious, be prepared for them to fight a defensive war; if he or she is highly analytical (type 3), anticipate their being slow in making up their minds.

If you're a competitor of the "new IBM," with its aggressive leadership, expect a fight. Everything about the forceful personality of John Akers, the chief executive, informs you that he and his company are type 1s—valiant fighters. Corporations *are* reflections of their leaders. If you know the man or woman, you know the company. For example, it has been said that when Frank Farwell left IBM to take over the presidency of Underwood Typewriter, he vowed that he wouldn't spend his life peddling typewriters and adding machines. His personal dislike could explain why Underwood shortly afterward dove into the computer industry totally without the expertise or finances to succeed in it.

Personality Factors in War

A general in the field should endeavor to discover in the chief that is against him, whether there be any weakness in his mind and character, through which he may be attacked with some advantage.

Polybius, 150 B.C.

Consistently, revenge is reported as one of the principal causes of war. "Retribution," wrote Ibsen, "is inexorable." The needs for power, achievement, and recognition also often supply the motivation for war, including business war. Each of these is a strong driver of corporate action. Below are some of the questions you might wish to ask yourself about your competitors' leaders.

- Who are the competition's leaders?

- What is their CEO like? Their key executives and managers?

- What beliefs do they hold dear?

- What is their background and experience?

- Who is rising fast in the executive ranks; what does he/she concentrate on that the company values?

- Are they satisfied with their company as it is, or are they dissatisfied and restless for change and new moves?

- Are they defense or offense oriented?

- If dissatisfied, given their personalities, what directions might they strike out in?

- Are they easily provoked?

- What is their chief flaw (impatient, too conservative, rash, slow, etc.)?

Tips on Understanding Your Opponent's Leaders

When waging business warfare, you must take into consideration the personality and personal preferences of the opponent's chief executive, his/her advisers, and other key managers. If you're opposing an invisible Howard Hughes-type opponent, you have problems, but otherwise there are reasonable methods you can use.

- *Draw inferences from written materials.* Great generals of history have had two habits that most managers never developed. They have generally been voracious readers and often were voluminous writers. Frequently they committed to books their values, philosophies of war, and strategies. Thus, Patton could read Rommel's book and anticipate what the field marshal would do. Most executives don't have much time for reading books, let alone writing them. But they may make public addresses and are interviewed by the press.

- *Recall observations.* At times even the most mundane piece of information can turn the tide of a great battle to your advantage. The decisive German maneuver at the Battle of Tannenberg was based entirely on the belief that the Russian general, Rennenkampf, would *not* march to aid his fellow Russian, General Samsonov. The German colonel who devised the maneuver had met the generals together before the war and had noticed their mutual enmity. At one point the two men had come to blows on a railway platform! The German simply inferred that neither man would help the other, no matter how much trouble he had fallen into. And he was right.

- *Test opponent's reactions.* If you know nothing about the person opposing you, it often proves worthwhile to try him out. Make a move and watch his reaction. This kind of thing is done commonly during any kind of negotiation. Whenever you sense that a move is being made solely to test your reaction, it's wise to follow the advice of samurai Asakura Norikage (1474–1555). "In time of battle or distress, the enemy may send out troops to upset a general in order to judge his feelings. The general should not show the least bit of weakness, however, or let out any word at all. The general should be very careful about this."

- *Ask around.* During the Atlanta campaign (1864), Union general Sherman learned that General Hood had replaced General Johnston as commander of the southern forces opposing him. Hood had been in the same class at West Point as two of Sherman's subordinates. Inquiring about Hood, Sherman learned that the man was bold, even reckless. Also realizing that Johnston had been replaced because he had not been able to stop Sherman's advance, Sherman expected a more determined effort by Hood. He thus moved his columns more slowly and cautiously against Hood than he would have had he been facing Johnston. Ask around about Thomas G. Plaskett, American Airlines' senior vice-president of marketing and you'll probably hear that he canceled his family's skiing trip one Thanksgiving in order to fill the seats with paying customers. Couple this with knowing that his background is in finance and you'll correctly infer that here's an efficient man who's concerned with economizing, even on nickels and dimes.

- *Prepare critical incidents.* A critical incident is a detailed report on the leader's decisions and behavior during a particular corporate episode. Combined with other critical incidents and other pieces of information, it can be used as a basis for inferring how he/she might behave in similar situations in the future.

- *Prepare a biography.* Increasingly popular is the "human-resource audit," a scouting report on the competitor's personnel. Focus particularly on the big achievements that have marked the person's career, his/her style of managing, and demonstrated attitudes toward risk. Does he select strategies with the most predictable outcome, or is he willing to pursue a big winner even if it means the chance of incurring a substantial loss?

THE OTHER SIDE OF THE COIN

As the term is ordinarily used, "security" consists of two elements: (1) compiling and analyzing information about competitors, which we have already looked at, and (2) preventing competitors from gaining information on you that could be put to their advantage against you.

If you're concerned enough to gather competitive intelligence about competitors, you can assume that they're equally concerned about you. In the same way you dream about having their new-product and marketing plans, their customer lists, product studies and projections and assessment reports, production schedules, merger plans, etc., so they fantasize about being in possession of yours.

The main source of information available to your competitor is your own staff—the salesperson who inadvertently reveals a company secret after having one drink too many, for example, or, far more odious, the employee on the take. According to the U.S. Chamber of Commerce, in excess of 30 percent of all business failures are the result of the actions of dishonest employees! A recent issue of *Small Business Report* contained a checklist of measures that a company could take to plug leaks, including:

- "Scrambling" product serial numbers

- Maintaining an employee code of ethics

- Requiring employees with access to confidential information to sign statements indicating their awareness of the sensitive nature of the material

- Preparing a contingency plan against theft

- Revealing to outsiders only as much information as needed

- Restricting or eliminating company tours

- Putting a limit on the amount of information available to salespeople

- Properly disposing of trash from wastebaskets

- Limiting the number of copies of sensitive information

- Discouraging public discussions among employees

- Having departing employees agree in writing not to divulge company secrets to new employers

- Identifying dissatisfied employees

- Accessing sensitive material on a need-to-know basis

- Rotating job assignments

- Refraining from putting information potentially useful to a competitor in corporate newsletters or on bulletin boards

- Making use of physical plant or office security measures

- Intensifying screening of prospective employees

KNOW THE CONSUMER

Knowledge of the country where he must wage war
serves as the base of all strategy.
Frederick the Great

An effective security system should surely compile competitive intelligence. But it will be incomplete without market intelligence that keeps you abreast of current and emerging needs of actual and prospective consumers. The company that doesn't know its consumers is never secure. That's why corporations are forever trying to uncover what makes a consumer buy.

Advertising's Leo Burnett Company's "Moment by Moment" system is a research method by which a panel of consumers watch television commercials and electronically indicate points where they experienced an emotional reaction, enabling advertisers to pinpoint events in the ad that impact on consumers.

Perception Research Services, Inc., of Englewood Cliffs, New Jersey, uses infrared light and cameras to track the movements of consumers' eyes as they scan magazine ads. Eye tracking has uncovered some interesting findings—that magazine readers notice only six out of ten brand names in ads, that photos that are powerful may cause the reader to overlook the product name, and that consumers are not necessarily more likely to notice more ads placed in the front of magazines as is commonly believed.

One study of consumer moods conducted by a member of New York University's marketing faculty found that consumers who are in a bad mood respond more favorably to cookie ads than those consumers in a cheerful mood but that most products receive a better response when the consumer is feeling good.

"We didn't realize the world was changing. We had to catch up," says John Richman, chairman of Dart & Kraft, at the time the parent company of Tupperware. As a result, Tupperware has added to its traditional marketing strategy of parties for homemakers. Currently, Tupperware also holds its parties in offices, before work, during lunch hour, early in the day or late—the parties fit the customers' schedule. And there are Tupperware parties for working couples or unmarried men and for the affluent or the less well off.

"Knowing the consumer." Some people in business do; many don't. I once bumped into a salesman who told me a humorous but true story of a car dealer who did. The salesman sells specialty advertising. His company will print whatever you want on any object—from calendars to pens, coffee mugs, even bowling balls. One afternoon he stopped at a used-car dealership. The owner listened to the salesman's pitch for just a few minutes, and then said, "I've got it. Here's what we'll do. I'll buy a bunch of your lighters and put an ad in the paper —BUY A CAR; GET A LIGHTER FREE. What a great promotion."

The salesman winced. Obviously the dealer didn't understand buying motives. He tried to put it as nicely as he could. "I don't know, Mr. ——. I really doubt people are going to buy a car just to get a free lighter. The lighters are good, don't get me wrong. But they're only worth a couple of dollars."

The dealer wouldn't be dissuaded. The salesman took the order, the lighters were delivered, the ad appeared in the paper—BUY A CAR; GET A LIGHTER FREE. You guessed it. As a result of the ad the dealer sold more cars than ever before in his fifteen years of being in business.

Knowing the customer is what automakers are trying to do with their experimental "personality cars," also known as "chameleon cars." The cars are so designed that they can be adjusted quickly to suit the personal tastes of the person driving them. For example, simply by flicking a switch, the driver can change the car's suspension to give a spunky sports-car feel or the sense of a heavier luxury car; or the transmission can be changed to shift smoothly or with a pronounced jerk. Mitsubishi's Galant gives the driver the ability to alter shock absorbers to a hard or soft ride.

In 1984, the National Institutes of Health dubbed osteoporosis a serious health problem. Weakening bones, the disease causes approximately 1.3 million fractures each year, particularly among women. Since calcium deficiency is one of the causes of the disease, the NIH recommended that women consume more calcium. In 1985, further research indicated that calcium might reduce high blood pressure and colon cancer. Companies making products that contained calcium quickly advertised the fact, and other firms added calcium to current products. Heinz changed the name of its Alba cocoa mix to Alba High Calcium and saw its sales outdistance projections by 50 percent. Tums are tablets of calcium carbonate. In 1985, shortly after Tums began running ads depicting its tablets as rich in calcium, its sales rose by 50 percent to almost fifty-five million dollars. After running second to Rolaids for three decades, Tums passed it almost immediately. Coca-Cola announced that it would market a calcium-added Tab diet cola, and Borden's declared that it would add calcium to its milk.

Coopers & Lybrand, the Big Eight accounting firm, started its EntreForum after looking at the unshakable facts and studying the signs. Small Business Administration reports reveal that between 1974 and 1984 the number of self-employed women in the United States grew 74 percent, from 1.5 million to 2.6 million. In the same period, the number of small businesses opened by males rose by only 24 percent in comparison. Coopers & Lybrand's EntreForum division links women entrepreneurs with bankers, attorneys, and accountants who are particularly aware of their needs.

The first step in winning is "seeing" the battlefield. This insight may seem obvious, but even presidents have been known to neglect to make certain that they were in fact seeing the battlefield. For example, one of the major causes of the Bay of Pigs fiasco was that neither President Kennedy nor any of the men advising him really studied a map of Cuba. Had they, they would have realized that the Cuban swamps and jungle would have made their invasion plan impossible.

Nowadays there are two very different schools of thought as to the answer to the question "What is the business battlefield, anyway?" The first school answers that the battlefield is your company's competitors. Those who subscribe to that point of view believe that the idea of business warfare is to defeat competing companies. In chapter 3 they were called great battlers whose "one-pole" strategy is to bring the competition to ground and defeat him. The extreme example of the great-battler company would be one that was totally obsessed with outcompeting a corporate rival.

The second school of thought answers the question "What is the business battlefield, anyway?" with "The customer is." Managers who adhere to this view are aware of the competition but not obsessed with it. These are "two-pole" strategists who can battle with competitors when that's called for but who prefer the second pole—maneuvering toward the consumer. They believe that corporations that are so wrapped up in competing are missing the point. In their view the idea of corporate effort is not to set out to do battle with other companies but to "get close to the customer"—to understand what customers like about the company's products and what they don't like and to give them the former and avoid the latter. Chapter 4 likened this approach to business warfare to the kind of war practiced by the eighteenth-century masters of the military arts, whose objective was not necessarily to do battle but to gain terrain.

The business-warfare counterpart of the military fight for terrain is the battle for the consumer. It's the most important corporate competitive struggle of all. In business warfare, the consumer is the real object of all your strategic and tactical maneuvers.

All we need do is substitute "knowledge of the consumer" for "knowledge of the country" and read "market segment" for "country" to see that Frederick the Great has a great deal to say about corporate marketing.

> Knowledge of the country is to a general what a rifle is to an infantryman and what the rules of arithmetic are to a geometrician. If he does not know the country he will do nothing but make gross mistakes. Without this knowledge his projects, be they otherwise admirable, become ridiculous and often impracticable. Therefore study the country where you are going to act!

Study the country where you're going to act, says Frederick. Numerous Japanese corporations are doing just that, often discovering that a hot item in Tokyo gets the cold shoulder in New York and vice-versa. Honda's small CRX two-seater gets up to fifty miles a gallon and looks classy. It's a big hit with young American professionals but hasn't found much appeal in Japan. American managers of Honda, more aware of American tastes, designed the CRX.

Other Japanese manufacturers are enlisting Americans to advise them on American tastes. Nippon Light Metal Corporation, for example, developed an ice-cream maker that was a successful seller in

Japan but failed when the company marketed it in the United States. It tried again, this time in a joint venture with American businessman James Kabler. One of the first modifications Kabler suggested was to change the size of the ice-cream maker for Americans. Originally the product came in a pint size only. That might be right for the Japanese, who customarily eat ice cream in small portions, Kabler thought. But when Americans eat ice cream, they eat ice cream. The American model was changed to quart size. The one-pinter had failed in America; more than four hundred thousand of the one-quarters were quickly sold in less than one year.

What's in a name? Doesn't a Coke by any other name taste as good? Maybe, but it doesn't necessarily sell as well. When Coca-Cola first began selling in China, it used a four-Chinese-character phonetic translation of Coke—"*Ke Kou Ke La.*" But as soon as Coke dropped that and changed the product's name to "*Ko Kou Ko Le*" sales quickly took off. Why? While the two sound alike to us, to anyone who understands the mind of the Chinese customer they mean quite different things. The original version means "Bite the wax tadpole." That's repugnant in any language. You can see why the second name generated sales. It means, "May your mouth rejoice."

Two questions matter most in business. The first is "What is the consumer interested in buying?" The second is "What are you trying to sell?" Though they are two questions, they should have but one and the same answer. Whatever its size, the competitively superior company will know better than its less successful competitors what the consumer wants and will be better at supplying it.

Those two questions, easy to ask, are extremely hard to answer. Of all the products put on the market only two in ten make money. This means that corporations are wrong about their products 80 percent of the time. This face is even more startling when you consider that for every one product that is released into the marketplace, twenty-four others have been researched, proposed, developed, test marketed, and shown to be unprofitable. In other words, behind the mere two in ten that become money-makers are 248 other products

that were considered promising enough by knowledgeable professionals to be test marketed. And it's all very expensive, the average
cost to introduce a new product nationally being in the neighborhood
of one hundred million dollars.

Levi Strauss didn't know the consumer in 1980 when it manufactured a low-priced but quality men's suit. Specifically, the company
didn't know that Levi's image in the mind of the consumer as a jeans
producer would work against it.

Sony and Panasonic did know the consumer. Perceiving the shift
in consumer tastes toward high-quality portable and table-model color
television sets, they started producing them. American firms didn't
know the consumer and stuck with mass producing large consoles
that lacked the quality of the small Japanese TVs.

One of the most fickle and unpredictable consumers is a child. The
toy industry is open territory for more than seven hundred manufacturers and their distributors, who have just about no idea what will
sell and what won't. It's a business driven by the blockbuster. The
recording industry has its gold and platinum sales plateaus; toymakers are all after the "megahit," the product that achieves sales of
at least three hundred million dollars within two years. But even
those most experienced in the business cannot predict with confidence
what will sell and what won't. Stephen Hassenfeld is head of Hasbro,
the largest toy maker in the world with $1.2 billion in sales. One of
the best in the business at selecting toys that will become big sellers,
Hassenfeld admits that his choices are usually based on intuition, not
market research. "They (Hasbro) do little or no research," says a
former competitor. "It's how everything feels in the gut." In 1984
the Michael Jackson doll was expected to explode and bombed. The
husky Masters of the Universe doll was thought to have a moderate
chance of decent success, but for one season. It became Mattel's big
item for three years and is still going strong.

Why is it that many corporations, even major firms making use of
ever-more sophisticated marketing-research methodologies, are so
wrong so often about what the consumer wants? One answer is to

claim that consumers are essentially screwy, their preferences elusive, ever changing, and like the wings of a live fly, very difficult to pin down. It's easy to defend this position.

Information Resources Incorporated, a Chicago marketing firm, developed sophisticated research techniques producing "single-source data" on how consumers behave. Single-source data are revolution-izing advertising and sometimes uncovering factual information that's yet to be explained. For example, Campbell Soups has learned that viewers of the soap opera *Search for Tomorrow* buy 22 percent less V-8 vegetable juice than the average, while fans of *All My Children* buy 46 percent more V-8 than the average. George Mahrlig, Campbell's director of media services, says Campbell has been "try-ing to unlock this thing, but it's still a big mystery."

Turtle Wax Incorporated uncovered a consumer oddity and turned it into a marketing venture. According to Chuck Tornabene, Turtle Wax's vice-president of marketing, "Every bit of research we've done shows not only a high consumer awareness of our name but a per-ception that we've been selling household products for some time." In fact, however, despite what consumers believed, Turtle Wax wasn't in the household products business at all but in auto-care products. It was only *after* the research had discovered the public's perception that the company decided actually to advance boldly into that business. As a result of the research, Turtle Wax developed five new household cleaning products in aerosol containers. It believes that it has a good chance of becoming one of the three best-selling brands in the seventy-five-million-dollar-a-year carpet-cleaning in-dustry. Turtle Wax took its cue from the consumer even though the consumer held a false impression of the company.

"Economic man" theorists tell us that if other factors such as quality are held constant, the consumer will select the product that costs the least. Homemakers in particular are thought to be particu-larly "economic" in their choices. Yet research studies often show that less than half the housewives had even looked at the price of the product they had just bought. Some people never check price.

Further complicating the view of economic-man and woman is that consumers will consider the price of one product but not another. One study demonstrated that shoppers were three times more likely to check the price tag on a detergent as on a box of cereal. Asked to state the price of cereals, 65 percent of consumers will not even be remotely correct, but 90 percent will be right if asked the price of a six-pack of Coca-Cola.

There are other schools of thought as to why consumers buy one product over another, including the "rational man" school and the "psychoanalytic" school. The former claims consumers compare products and select the one best fitting their needs; the latter school looks to irrational, unconscious forces to explain buying preferences. The same kind of exceptions, inconsistencies, and downright contradictions we saw when looking at "economic man" could also be cited to challenge those other schools, too. We should probably add "aesthetic man" and "social man" to this list, too. Behaving like rational man, consumers sensitive to red-meat health concerns have starting eating more fish. The consumption of catfish in this country has quadrupled since 1980. Display trout filets and catfish filets side by side and the trout will outsell the catfish, but only by three to two. But display whole trout next to a whole catfish in a grocery cooler and the trout will outsell the catfish by three to one. It's just not aesthetically pleasing to look at a catfish's face.

Aesthetic tastes are also affecting the (pardon the pun) growing "designer vegetable" business. Seed companies involved in the $1.3-billion-a-year home-gardening industry are marketing blue potatoes, white beets, white carrots, purple beans, and "inside-out" radishes (red on the inside.)

In 1983 Baycliff Company began marketing a line of Japanese food products. Packaging designers told Baycliff that the items wouldn't sell because the packaging was black in color. But the first distributor Baycliff approached liked black. So did consumers. Once thought by marketers to be absolute anathema to product sales, black is currently the in color. In a recent Package Design Council's Gold

Awards contest, seven of the fourteen award winners were black. Everything from GM automobiles to Heinz vinegar is now being packaged in black.

That many people are sociable—social man—is one explanation for the difficulties manufacturers of banking ATMs—automated teller machines—fell into. Reportedly sales of ATMs declined 29 percent in 1985 and 16 percent the year before. Less than half of bank customers have used them, and only one-third of that half uses them regularly. Many people say they don't like them because they prefer dealing with live tellers.

Nabisco relied on the concept of "patriotic man" to help sell its Almost Home Cookies. To induce consumers to try that new product, Nabisco offered to send containers of the cookies to men and women in the armed services if the family mailed in three proof-of-purchase seals. "The soldier angle," said one corporate manager, "makes people like us and remain loyal." In 1983 American Express also appealed to the patriotic man in consumers when it donated one cent to the Statue of Liberty restoration for each transaction in which its credit card was used. The number of transactions increased by almost one-third. Branching out from patriotic man to cause-oriented man, American Express has connected itself with fifty causes, including donations to a fund for the preservation of Norway's national bird.

TV show syndicators are finding profit in "nostalgic man," discovering that now middle-aged baby boomers enjoy the same shows they liked in their youth, from "Leave it to Beaver" and "The Honeymooners" to "Combat," "Gidget," and "The Groucho Marx Show." Networks are capitalizing on nostalgia, too, by offering new productions of old shows such as CBS's "The Man from U.N.C.L.E." Turner Broadcasting aired "The New Leave It to Beaver" in fall 1986, offering that show on Monday nights and the 1957–63 original "Beaver" on Monday through Thursday nights.

Another answer to the question "Why are corporations so often wrong about what consumers want?" is that market-research methods, however sophisticated, are limited. Even when market analyses demonstrate strong consumer interest in a product, the company that brings out the product is still taking a gamble. One of GM's executive engineers hit the nail on the head when he said with the simplicity of the experienced businessman, "I don't think consumers can project themselves into the future very well to say what they will want." The major limitation of research in piercing the fog of marketing and product-development war is that it deals only with "revealed preference"—the preference of the consumer as revealed by what he or she buys. It doesn't identify whether or not they will buy a product that hasn't been invented yet.

On the Spot: Taking a Look, Sniffing the Air and Listening

Vegetius's *Military Institutions of the Romans* was one of the most influential books on warfare ever written. In it Vegetius makes the simple observation that the greatest generals were not satisfied with the descriptions of the country in which they were engaged but laid their plans "on the spot" itself. Napoleon was an on-the-spot commander. He said, "A general who has to see things through other people's eyes will never be able to command an army as it should be commanded."

Here he is shortly before the Battle of Austerlitz, one of his most remarkable victories. Eight days before the battle, he mounts his horse and rides out to survey the area, accompanied by his staff. He takes note of a number of features of the terrain—the low, soggy ground holding a few meager villages that could serve as a temporary defense line; the northern end of the valley that cannot be seen by the enemy's observers and could be used as a site-of-force concentration; the high ground held by the enemy, the Austro-Russian army, that from a distance looked precipitous but close up reveals two slopes that could be easily ascended by his troops. His eyes also take in the Santon, a small, swelling hill. He points to it, tells his entourage that a battle will take place there, and orders his engineers and gunners to begin immediately to prepare it for defensive action.

He is proved correct eight days later when, after the ebb and flow of forces, the French army rolls into position precisely at that point.

What's most instructive about this incident is that Napoleon, the greatest warrior of modern times, *went out to take a look for himself.* He rode around, sniffed the air, looked into the valley and up the hills. He had maps in his tent, and piled next to them he had reports from his officers, and he had the verbal reports of his staff. But still he mounted his horse to take a look around.

Comedian Steve Martin says, "Reality. What a concept," and brings down the house. Much would be gained if every once in a while someone in a corporate meeting were to say loudly, so everyone could hear, "Reality. What a concept," reminding everyone huddled around the table that reality, after all, is what's being searched for. It's very easy to forget that maps and written and verbal reports, like marketing surveys, printouts, and schools of thought on why buyers do or don't buy are all symbols of reality and not reality itself. They aren't just once removed from reality but twice removed, for they are symbols of someone's interpretation of reality. To read about the lay of the land is not the same as to see the lay of the land or to smell it, for yourself.

I was reminded of the importance of "sniffing the air" myself some years ago. I was heading a large team of consultants whose job was to advise a government agency on where specifically to locate its fifty or so offices in a large metropolitan area. After six months of intensive, in-office analysis of every type of demographic, performance, transportation, and census data imaginable, we believed we had a pretty clear idea of the best possible locations for offices. I felt a little uneasy, however, and had all the team members fan out to all our proposed sites and telephone me from a public phone. Often I learned more from one of my staff just telling me what he or she *saw* from the booth than from all the reports. And when I went out and saw for myself, I learned even more. The data had helped; the sniffing helped even more.

In *A Passion for Excellence,* Tom Peters states that many accused

In Search of Excellence, which he coauthored with Robert Waterman, of oversimplifying. Peters reached the opposite conclusion: the book didn't simplify enough. In Peters's view, when all is said and done, the number-one productivity problem in the United States is managers who are out of touch—out of touch with their people and out of touch with their customers. He adds that the alternative—being in touch—doesn't come from reading printouts; it means exploring and interacting, not issuing or reading reports.

In my mind, "being in touch" is always related to the "hands test" of A. D. Moore, an expert on creativity. Moore claimed that when a person is trying to be creative, he will automatically want to get his hands into the act. He'll want to fiddle with the thing, hold it, turn it around, flip it upside down.

Moore contrasted the "hands test" with merely reasoning about a thing. He pointed out that we overestimate the contribution of the Greeks to science and overlook their many conclusions that were out-and-out ludicrous. Moore claimed that the Greeks went wrong simply because they didn't do any really practical work. Their slaves did that. The Greeks didn't have to use their hands and considered doing so beneath their dignity. As a result, "handless," the Greeks relied on hypotheses and reasoning to reach their conclusions.

The "hands-test" manager will want to bring his hands into the job of learning what the consumer wants—and his ears and eyes. As Vegetius said, and Napoleon illustrated, he'll make use of reports addressed to his reasoning powers, but he won't rely on them to the exclusion of what his senses tell him directly.

Some of the world's most successful corporations employ their own versions of the hands test. Every one of Wal-Mart's executives is required to spend at least one week every year working in a local store—not managing, not giving orders, but waiting on customers and unloading cartons. McDonald's runs more than two thousand restaurants outside America. Long before the company locates an outlet, it has its executives walk through local grocery stores to gauge the area's supply of hamburger buns and condiments.

Pepsi-owned Frito-Lay, the leading salty snack company in the United States, is legendary for its insatiable appetite to stay in close touch with the customer. Every week the 9,400-person Frito-Lay sales force makes almost three-quarters of a million calls on its 325,000 retail customers, providing a number of services, including fluffing up Frito-Lay packages so they look more appealing to the shopper. To facilitate order taking and help it track sales trends quickly, the company is investing forty-five million dollars in hand-sized computers for its sales people.

Procter & Gamble is often credited with knowing its customer better than any other household products corporation in the United States. Pillsbury, a competitor, claims that P&G has lost its advantage in consumer intelligence because the customer P&G knew better than anyone—the traditional American housewife—no longer exists. Today's homemakers, researchers say, are far different from their mothers. They cook less, clean less, are less enamored of quality but more interested in convenience. And they are more than ever before influenced by the preferences of the husbands and children. In one study, when husbands and wives were put in a supermarket together, the men picked different brands than their spouses 43 percent of the time. Teenage daughters chose differently even more—half the time. Yet children and husbands are doing more of the shopping nowadays.

Procter & Gamble believes it's still first in customer knowledge. Said one P&G spokesman, "We believe our strength always has been knowing the consumer well, no matter who the consumer is." P&G was among the first in the field to advertise directly to men; and to see for itself how the "new woman" actually behaves, P&G has had television cameras mounted in kitchens and other rooms. One thing it noticed is that while in the past householders filled the sink with soapy water when dishwashing, today's consumer just squirts the dishwashing liquid right onto the utensil or plate. A less highly concentrated dishwashing liquid may be called for.

Between 1982 and 1985 GM's Pontiac division upped its sales by 64 percent. Even while the other GM divisions and Chrysler, Ford, and American Motors were experiencing sales declines in the first half of 1986, Pontiac's sales rose by almost 10 percent. Just one of the numerous methods Pontiac used to understand what consumers wished to buy involved hiring a psychologist to ask consumers to describe their fantasies about driving.

Kelly Services ("Kelly Girls") invented the temporary-office-help business as a result of listening to its customers. Originally, founder Russell Kelly's idea was to bring the typing and duplicating into his offices, but when consumers started requesting temporary help in *their* offices, Kelly quickly changed his business from service assistance to temporary help.

The Craftsman's Orientation

In business warfare the person who knows the consumer terrain better than anyone else is the very one who nowadays gets the least "ink" in the press. It's the small craftsman who, like Napoleon, goes out to see for himself and who, unlike the Greeks, is all hands. He spends considerable time with the buyer, and jointly they carefully forge the precise specifications of quality and price. He listens closely to the customer, checks back, delivers on time, gauges satisfaction, checks again. He individualizes, treating each customer as a unique person, not a generic segment that can be pigeonholed and stereotyped. Above all else, the craftsman stands for quality. He may or may not have heard of "corporate total quality control," but he practices it day in, day out. Like the guerrilla he is, he knows that a quality product or service wins the consumer's loyalty to him and that shoddy workmanship loses it. He is aware that the cliché that a satisfied customer is the best advertisement is really true and that an unhappy one is the worst advertisement. If he makes one customer happy, that one tells ten others; but if he makes one unhappy, twenty others hear about it.

All of the recent emphasis on

- warm-armpit marketing

- close to the customer

- putting your 800 toll-free number on your product (and, like Procter & Gamble, finding that the 800 number is a major source of ideas for improving products)

- trusting experience and intuition at least as much as marketing research, and perhaps more . . .

. . . are various ways of saying that no matter how large your company and how standardized your product, you'll fare best in your competitive wars if you and your staff act as if you're all small craftsmen. "The course of war is hidden," said King Archidamus of Sparta in the fifth century B.C., "and much comes about from very small things"—like simple person-to-person conversations with customers.

Guard Against Too Much Monkey See, Monkey Do

You and I have seen it many times. One person asks another, "What time do you have?" The second checks his watch and says, for instance, "Ten thirty-two." Then the first adjusts his watch to ten thirty-two. If there are other people within earshot, they usually adjust their watches also. Rather than thinking, *My watch is accurate; it says ten twenty-nine. I'm right; he's wrong*, they think, *Oh, mine must be wrong*. Obviously this is justifiable if the competitor's product, "his time," is the correct one and yours isn't. The problem is when the competitor's time is wrong.

Corporate managers all too frequently assume that if the competitor —particularly the leader—is doing something different than they are, the competitor must be right and they must be wrong. So they play follow the leader. For example, in just one three-month period between April and June 1986, 87 percent of the packaged products introduced into the marketplace were "me, too" goods—almost nine out of every ten products, many of them developed by the nation's best marketing companies, the very ones that swear that the safest path to business success is by way of a better mousetrap.

In 1815, financier Nathan Rothschild used his competitors' monkey see, monkey do tendency against them, using what espionage agents know as "phony revealments." Rothschild knew that his competitors closely watched his every move. He also knew before they did that the British had won the Battle of Waterloo. His intelligence system was superb. He then sold his British government securities, deflating the market. His competitors, concluding that he had inside information that the English had lost, sold theirs. Suddenly Rothschild bought them back at an absurdly low price. When the news of victory came, the value of the securities naturally skyrocketed, and Rothschild had made another fortune. Even today "phony revealments" are still being cashed in on. For example, it's not impossible for futures traders to leak phony information on key indicators and to profit from the market reaction.

The heated bomber-versus-battleship controversy earlier this century points up how tempting it is to do what everyone else is doing merely because they're doing it. In July 1921, the U.S. Army Air Corps sank the supposedly "unsinkable" ex-German battleship *Ostfriesland* in tests off the Virginia Capes by dropping a few heavy bombs from planes. This was the first actual demonstration that gravity bombs could give a ship with heavy armor the deep six. Yet, in spite of the evidence, American commissions appointed by Presidents Coolidge, Hoover, and Roosevelt and British boards of investigation all came to the conclusion that as long as other nations were building battleships, America and Britain would have to follow suit.

Corporations have been known to do the same thing, even enormously large ones. In the late sixties Standard Oil of Ohio's (Sohio's) industry-analyses department recommended to upper management that the company diversify into the nuclear-power business. The analysts didn't base their recommendation on direct evidence of emerging energy needs of consumers but on a survey that indicated

that Sohio's major domestic competitors were becoming heavily involved in the nuclear industry. In the mid-seventies, Standard Oil of
California based its move into nuclear fuel at least partially on an
internal report that pointed to *rumors* that Exxon had been proceed-
ing on the basis that the price of uranium would increase drastically.
The report also cited Exxon's hiring more geologists and recruiting
100 summer students for jobs in the minerals field.

American companies often wait and watch while another corporation takes the risks of testing a new product. If the product
succeeds, then the competitors rush on to the field of battle. Japanese
companies, too, often employ "monkey see, monkey do" to good
advantage. When entering the United States, they often contract with
American competitors to distribute their products and establish them
in American markets. After learning just how the American firms
operate, the Japanese often create their own sales-and-distribution
network and compete directly with the firm that they used to establish themselves. However, total monkey see, monkey do means getting
so engrossed in emulating what the competition is doing that you
fail to see what the consumer wants. It's basing your strategy solely
on someone else's judgment rather than on your own independent
analysis.

Number two in any industry is often wrongly seduced into attempting to copy what number one in the industry is doing. Wal-Mart, the
number-two retailer, is successful precisely because it plays its own
game, not that of its chief competitor, K mart. Wal-Mart is aware
that its market is small-town America, which it knows like the back
of its corporate hand. And that is where it stays and thrives. As a
direct result of refusing to play monkey see, monkey do with K mart,
Wal-Mart is the fastest-growing retailer in the United States. Fisher-
Price adamantly refuses to follow a monkey see, monkey do course
and is the largest manufacturer of infant and preschool toys. More
important to the company than what the competition is doing is
what its consumers say they want. On a regular basis Fisher-Price

invites groups of three-year-olds to its East Aurora, New York, head-quarters to pass judgment on new product possibilities. If the kids don't like a product, it's not marketed, no matter what competitors are doing.

Be Careful About the Other Form of Madness

The other side of monkey see, monkey do is monkey see, monkey too scared to do. That is, after uncovering what the other firm is about to do, corporate people in staff positions, like their military counter-parts, frequently refrain from recommending the same or similar moves because they give the other firm credit for possessing capa-bilities that that firm in reality doesn't possess. The threat of competi-tion is enough to scare many firms off even though by competing they would win.

Beware of Groupthink

If you're like most managers, you spend between 65 and 80 percent of your time with one or more people in meetings. It's through groups meeting that all a manager's main functions are performed—plans laid, units organized and staffed, efforts coordinated, reporting achieved, and budgets hammered out. Groups diagnose the problems and the need for decisions; groups develop the solutions and deci-sions, and groups mobilize the action to implement the solution or decision. And it's through groups that corporations make all the key decisions about what the consumer is interested in buying and what you're trying to sell. If a corporation consistently miscalculates, it quite possibly is because of what happens or doesn't happen in its meetings.

It's generally considered useful to appoint teams of executives representing major parts of the firm to formulate a firm's strategy. However, Prince Eugene and Napoleon both used to say that the only reason for calling a council of war was if the general had no desire to fight, for in such councils the decision to fight or not was generally not to.

Eugene and Napoleon were leery of what corporate strategic teams must always be cautious to avoid—"groupthink." It's "a deterioration of mental efficiency, reality testing, and moral judgment that results from in-group pressures." In biology, inbreeding often results in the development of unpleasant recessive traits. Groupthink does the same in corporations. Groupthink is another reason why companies have difficulty understanding the needs and preferences of consumers. In the book *Victims of Groupthink*, Irving L. Janis identifies six major defects in decision making that contribute to the flaws of groupthink:

1. The group's discussions are limited to only a few courses of action, usually only two—"We could do X, or we might try Y."

2. The group fails to reexamine a course of action favored by the majority from the viewpoint of risks or disadvantages that weren't originally considered.

3. A course of action once rejected by the majority tends to remain that way; it isn't reconsidered.

4. Group members make little or no effort to solicit the views of experts on the subject.

5. Members discuss information that supports their opinion and ignore information that doesn't support it.

6. The group spends little time developing contingency plans to achieve the objective that could be used in the event things go wrong.

What encourages groupthink? First, group cohesiveness and loyalty, which, ironically, is what most corporate CEOs try to promote, as in "Harry is one of the team; he fits us like a glove." Few qualities are valued more highly in corporations than loyalty. It's a kind of corporate morality. In the positive sense it breeds team spirit; but when carried to an extreme, it reveals itself in staying with the decision made by the group even if the decision is a bad one. Second, isolation of the decision makers from the opinions of qualified associates who, not members of the group, are outsiders. The majority of corporations use participative methods to improve communication, overcome resistance to change, increase employee satisfaction, etc. But the main reported reason is to enhance the quality of the decisions. Do groups produce decisions that are better than the decisions that one of its members could make? Supposing that for any decision there are right answers and wrong ones (that, specifically, there are right answers to the question "What does the consumer want?"), the situation is this: if members of the group are wrong but *confident*, the group decision tends to be wrong; but if the members of the group having the right answer are also confident, the decision of the group tends to be right in the great majority of situations. Keeping the issue open until judgments of outsiders are expressed is preferable to merely informing them what the decision is after it has already been made. This means avoiding the "chiseled in stone" decision.

A third factor contributing to groupthink is the leader who promotes his or her own preferred view. We're not talking about the stereotyped corporate leader who turns his advisers into robotlike yes-men but of the more subtle influences the leader impresses on the group that even he or she might not be aware of. For example, merely by asking for Tom's opinion more often than anyone else's, the leader may inadvertently be communicating that he literally likes to hear what Tom has to say. Only a foolish manager will not take note and shift a little more to Tom's side.

How can groupthink be avoided?

- Valuing loyalty, but also open, critical dialogue.

- The leader being impartial and not stating his preferences at the beginnings of deliberation; expressing his views, but not pushing them.

- Involving other individuals and advisory groups and subgroups in discussion.

- Being willing to alter a decision even if it means a major change of direction. Getting across the idea that no decision is chiseled in stone.

- Assigning one person at each major decision-making meeting the role of the devil's advocate.

WHEN ON THE DEFENSIVE

The best armor is staying out of gun-shot.
Italian proverb

Any corporation can be attacked at any time—yours, mine, and everyone else's. Defense isn't a principle of war, as we saw in chapter 5. Offensive action is. In war of any type the best defense is an aggressive offense. If you remain on the defensive, you're toying with defeat. But you will be hard to beat if you're relentlessly taking the offensive to lower your costs and to concentrate your forces along lines that differentiate you from your competition along his lines of least resistance.

But even the most aggressive companies like PepsiCo and corporate leaders—Victor Kiam, for example—cannot *always* be on the offensive. In war there has never been an uninterrupted offensive. In all of history, few men have been addicted to offensive action as thoroughly as World War I general Ferdinand Foch. But even he had to turn to the defensive when the offensive was out of the question.

Foch was an extraordinary charismatic man who is best remembered for the contagious power of his personality to inspire anyone who came in contact with him, his unflagging faith in offensive action, and his unswerving belief that in war the side with the greater spiritual strength would win. A battle is lost, he said, when "one *believes* one has lost." A battle won "is a battle in which one will not confess oneself beaten." Battle equals a struggle between two wills; victory equals will.

Foch's own moral fiber exemplified his aggressive belief in the power of the forceful human spirit over technology or any other element of war. An associate draws a portrait of Foch in action as the supreme commander of Allied forces in 1918:

> bursting like a whirlwind into every headquarter, his face contracted, his body all tense and contracted, gesticulating, fulminating in jerky ejaculations. To a general in agony who tells him, "My troops are yielding under superior numbers; if I do not get reinforcements, I cannot answer for anything." He replies . . . "Attack," the general will insist. "Attack, attack, attack!" bellows Foch. . . .

To a supreme Allied council, some of whose members are contemplating retreat, he begins by saying, "You are not fighting. I want to fight. I would fight in front of Amiens. I would fight behind Amiens. I would fight all the time."

When he took command of all operations on the Western Front in March 1918, the situation was such that Foch could not follow his personal preference for attack. His only strategic course of action was to take the defensive and hold out against the hard-hitting German offensives. For two months the French held on, barely. The Germans struck a sudden, forceful blow on May 27 at the Chemin des Dames, taking the French by surprise and opening a wide breach. For the first time in four years the Germans poured through, covering ten miles a day, reaching the Marne three days later. Another German attack was delivered on July 15, but this time Foch was ready, having anticipated their blow. He brought up his reserves and launched his first counteroffensive. On July 24, with the arrival of fresh American troops, Foch declared that the tide of battle had turned: "The moment has come to abandon the general defensive attitude forced upon us until now . . . and to pass to the offensive." For the next three months Foch fought everywhere. He never let the enemy rest. Offensive followed offensive at one point, then another and still another. The Germans were driven back and soon beaten.

In the long run it's always preferable to take the offensive, but at certain times doing so is impossible. Every corporation, like every army, is sometimes forced to take the defensive because, as in football, the competitor has the ball and is moving down the field. You take the defensive when you lack the strength to attack or when your competition has dealt you a blow. At those times, just like Foch, however much you would prefer to "fight all the time," you can't. Instead, you must fortify yourself and struggle to protect every inch of marketing ground, every dollar, every part of your company that can be saved.

Defensive Tactics

You can undertake defensive operations for a number of reasons, including these:

- To dissuade competitors from attacking

- To defeat their attack

- To gain time

- To concentrate your manpower elsewhere

- To wear the competition down before you launch your offensive

- To maintain control of "terrain" that's important to you

- To force them to mass so they're more vulnerable to your fire.

The best defense is one that discourages your competitor from attacking at all. If you can keep the competition completely out of your hair and away from your markets, your defense is strong. If that's not possible, your defensive objectives shift to inducing competitors to channel their competitive attacks in directions that are less threatening to you or to force them to reduce the intensity of their attack. If a competitor is challenging you, you can discourage him by blocking him, threatening him with retaliation and making the prospect of attacking you less attractive to him.

You can discourage him by blocking him along the lines of his probable concentration. If the competition intends to concentrate along the lines of a core strength, you can make yourself superior to him in that strength. If the competitor is attempting to gain certain "terrain," you can increase your strength in that area. If he is attacking your vulnerability, you can shore up that weakness and make it a strength.

Threatening severe retaliation consists of letting the competition know that you won't sit idly by while he attacks. You'll gain a great deal by establishing a reputation as a vengeful retaliator. A competitor will think twice about going after what is now yours if he knows you'll counterattack with full force *every* time he tries it. Every neighbor of Israel knows that any action against that country will be met with an extreme reprisal. When PLO operations based in Lebanon blew up an Israeli civilian bus, Israel did not retaliate by blowing up one of theirs but by sending in its tanks and seizing half of Lebanon! The key factor is your consistency in retaliating. If competitors believe they'll feel the sting of your vengeance *every* time they attack you, you'll be much more secure than if you retaliate only periodically.

Puny Royal Crown demonstrated that you don't have to be a giant to retaliate with a vengeance. When Coca-Cola offered to buy Dr Pepper for $470 million and Pepsi to purchase the Seven-Up Company for $380 million early in 1986, Royal Crown first brought its charge of unfairness to the public's attention through newspaper ads. Then it filed an antitrust complaint with a district judge. On June 20, 1986, the Federal Trade Commission blocked both Coke's and Pepsi's acquisitions.

Another method of discouraging your competitors from challenging you is by making the prospect of attacking you less attractive to them. The promise of profit makes what is now yours appealing to a competitor. Greed is one of the chief flaws of a commander and company. By being greedy in a market or business, you increase its appeal to a competitor. But if you lower profits, reduce prices, and increase discounts, you make the prospect of attack less appealing to competitors.

Stick to the Active Defense

A strictly passive and permanent defense leads to no gain. All of World War I demonstrated the futility of the passive, entrenched defensive. You're reminded of what was said in chapter 5: stay on the defensive if you're forced to, but the moment the competition makes a bad mistake or has exhausted itself in attacks, turn quickly to the offensive, "the flashing sword of vengeance."

The defensive should be only a temporary position, never a permanent attitude. The only legitimate reason for staying on the defensive is because you aren't strong enough at this time to take the offensive. When you're prepared and strong, seize the initiative. In some instances a company can be on the defensive for a long time before making its offensive comeback. The A&P has been in the defensive mode in one form or another for a few decades, only recently showing signs of offensive resurgence. A&P and the American steel and oil industries have had to be on the defensive. But often corporations take the defensive that have no business doing so.

Why do so many firms that should be on the offensive cling to the defensive? There are many reasons certainly, and one of them is that defense is easier than offense. If your managers are run-of-the-mill, you'll tend to find yourself on the defense even when it's completely wrong. Average generals and managers never like to exert themselves more than absolutely necessary.

Always Be Skeptical

"Skepticism," wrote Frederick the Great, "is the mother of security." For the sake of security, always assume that some competitor or another is getting set to attack you. The one night an army becomes lax about its security is usually the night the enemy attacks; the single road left unguarded is the one he uses.

Corporate confidence and enthusiasm are positive qualities, but a sprinkling of realistic nervousness about the competition is very healthy. A dear friend of Frederick's, for example, claimed to be able to foretell success or failure from the great man's moods. Whenever Frederick showed too much confidence, a reversal was in store; when he showed fear, a success soon followed.

Don't just assume that the competition has designs on what you now consider yours; assume ambitiously. Give him credit for a touch more boldness and audacity than you want to think he really possesses. Put to use your intelligence on the other company and its leaders and estimate what they might try to attack. American firms competing with Japanese companies should generally think in terms of being attacked along two dimensions: smaller-size products and lower prices.

Maintain a Multifaceted Defense

Often even a vigilant company will assume that the competition will attack in a particular way—through acquisition or by modifying X product or by expanding into another region, etc—and will design its defensive system around these assumptions. Affording more security is the system that is *not* made to detect and meet any specific type of threat but leaves room for all eventualities.

Avoid a Maginot-Line Mentality

Fortresses are the tombs of armies.

Military maxim

Historically, permanent lines of defense have failed. The Maginot Line, Mannerheim Line, Siegfried Line, and Bar Lev Line were all overcome. The Mongols thought the Great Wall of China was a joke. Such "lines" are physical, tangible expressions of a passive attitude that eventually brings defeat, an attitude of aversion to the offensive.

Who establishes permanent lines of defense? It's always the rich and powerful state or firm. Guerrillas don't, and the most aggressive fighters don't. The Mongols, the most ferocious combatants in all of history, never surrounded their cities with walls. Probably most every company that once was great but fell from greatness did so because it remained too long and too passively on the defensive.

Mobile defenses are superior—flexible, elastic defenses based on the concept that if the competition penetrates your line, his flanks will be vulnerable to attack. Mobile defenses are predicated on the notion that from the get-go the defender plans to launch a counter-offensive as soon as the opportunity presents itself.

Not many years ago Coopers & Lybrand was the largest Big Eight accounting firm. Then Arthur Andersen & Company took the number-one slot. In early September 1986, Peat, Marwick, Mitchell & Company announced its intention to merge with Amsterdam-headquartered KMG Main Hurdman to create the Big Eight's biggest company, nearly twice the size of Andersen. Immediately, all the other firms were thrown on the defensive, but they made it clear that their defense would be mobile and active. "If Peat Marwick tries to take our best clients away, we'll take some of theirs in retaliation," some said. Others laid plans to pursue Peat Marwick clients during the corporate confusions that would be created as the two firms attempted to adjust their cultures to the merger.

Secure the Key Areas

He who tries to defend everything saves nothing.
Ferdinand Foch

ABC did a remarkable thing in programming for the 1986–87 television season. It conceded a half-hour prime-time slot every Thursday night to NBC. Rather than attempting to compete with NBC's *Cosby Show*—at the time the highest-rated show on television for two years running—ABC took a course of least resistance and put an inexpensive news show, *Our World*, in the time period. Some considered it admitting defeat. Clearly it was, but it was also a move that saved ABC about eight million dollars that it otherwise would have spent on a dramatic series that even after that money had been spent would not have cut into Cosby's viewership. The *Cosby Show* is the porcupine ABC had no intention of getting into a pissing contest with. When the show's popularity declines, that's when ABC will compete.

The ABC example illustrates an important lesson of business warfare when you're thrown on the defensive. It is impractical, indeed impossible, to be certain that your competition isn't stronger everywhere. Writes Ferdinand Foch:

> There are many fine generals, but they try to keep an eye on too many things; they try to see, to keep, to defend everything . . . such and such a strong position. Using such methods . . . in the end means dispersion, which prevents them from commanding, concentrating on one single affair, from striking hard; they end in impotence.

Recall what we said earlier. If your competition is willing to pay the price, they will be able to penetrate even the strongest defensive position—at least temporarily. The secret is to keep the price so high that not even a fool would be willing to pay it. Your defenses should occupy your important strategic points, not every point. Fortify yourself where you wish to prevent the competition from launching a successful attack—and fortify in mass, creating strong points that are prepared to withstand a direct assault. In chapter 4 we looked at concentration of forces for the offensive. Concentration also operates when you're on the defensive. The following are guidelines for defensive concentration:

- *Concentrate quickly.* The information on the competition is never complete. Yet in spite of this, you must concentrate immediately when he attacks.

- *Concentrate sufficient resources.* The rule of thumb for defensive concentration is "never allow your opponent to become three times stronger than you at the point of his concentration."

- *Take chances elsewhere.* To concentrate sufficient defense at point X means that you must temporarily weaken points Y and Z.

- *Bring up support.* To be successful, defensive concentration requires coordination of units and staff in addition to the one under attack. Fight as a combined team.

- *Beware of oozing through.* A competitor will sometimes precede a large-scale infiltration by oozing through small advance forces—taking just one of an advertising firm's accounts away, for example, or a beer company signing on a few small distributorships in a region held by a competitor. At times the first sign of oozing through is so insignificant as to be almost unnoticeable. The only defense against oozing is constant vigilance and bringing up your reserves quickly to stop the oozers immediately and decisively.

When in Retreat

It's axiomatic that after losing a battle you shouldn't fight a second battle unless fresh, favorable circumstances come into play. Retreat is continued to the point where an equilibrium of force is restored, as by way of consolidating in a strong defensive position, by recruiting reinforcements, or by a waning in the competitor's concentrated attack.

A slow retreat is preferable if possible, one that moves backward begrudgingly—the retreat of a wounded lion that, even as it moves, turns periodically and strikes back menacingly.

If ever your company is in retreat after being beaten, pay particular attention to the mental state of your personnel.

> When troops once realize their inferiority, they can no longer be depended on. If attacking, they refuse to advance. If defending, they abandon all hope of resistance. It is not the losses they have suffered but those they expect to suffer that affect them.
>
> Soldiers who have ceased to hope are no longer receptive beings.

Returning to the Offensive

The turn from the defensive to the offensive is what the defensive is designed to achieve. The topic of this chapter is security, and the height of security is gained by imposing your corporate will on the competition by seizing the initiative and maintaining it through one offensive action after another.

The ideal moment for a major counteroffensive is when the competition has fully committed itself to an attack and yet has not achieved the objective it desired. If you counterattack, then your chances of success will be substantial.

When the gains that can be achieved by counterattacking are greater than those of staying on the defense, counterattack. When the competition has captured critical terrain from you, counterattack to reclaim it.

In football you cannot have your defensive and offensive units on the field at the same time. In business warfare you must. Even when you're on the defensive, you must be planning for the shift to the offensive.

CHAPTER EIGHT

Making Certain All Personnel Play Their Part

It is not big armies that win battles; it is the good ones.

Maurice de Saxe, 1757

Bigness is not a sign of strength. In fact just the opposite is true.

Martin S. Davis, Chairman
Gulf & Western Industries, 1985

Dr. Benjamin Franklin (B. F.) Goodrich opened his rubber company in Akron, Ohio, on the last day of 1870. At the time, many of the rubber products being manufactured in the country were of poor quality. In addition, rubber wasn't being put to many uses. Goodrich decided to put his mind and tremendous personal energy to solving those two key problems: product quality and new-product development.

In the company's first year it developed its first major product— a cotton-covered rubber fire hose that was much superior to the leather hoses it replaced. In the same year it added other hoses, gasket rubber, jar rings, rolls for the hand wringers used by housewives, and bottle stoppers to its product line. To make certain that the products were of the highest possible quality, Goodrich personally supervised the manufacturing process. Having solved the product-development and quality problems to his satisfaction, Goodrich then turned his attention to the next problem—finding consumers to buy the company's fine products. He shifted the corporation's priority

283

from products to peddling, even to the extent of packing his bags and himself selling.

Goodrich was illustrating in microcosm what Napoleon called "the great art of war"—the art of the correct distribution of forces. It's putting your resources where they are most needed and will do the most good. When the key activity was product development, that's where Goodrich placed his emphasis; when sales were needed, he hit the road and sold. The leader should always send enough men or women to accomplish the mission, in Goodrich's case, enough of a man.

This competition within the competition of business warfare is to distribute your forces more advantageously than your competitor. The war principle is the principle of ECONOMY OF FORCE. Ferdinand Foch described it as "knowing how to expend, to expend usefully and profitably, to make the best possible use of all available resources" and doing it by "pouring *all* of one's resources at a given moment on one spot; of making use there of *all* troops."

Small corporations may be able to pour *all* of their resources at one spot, but larger, diversified corporations can't. More useful is that "an army should always be so distributed that its parts can aid each other and combine to produce the maximum possible concentration of force at one place, while the minimum force necessary is used elsewhere." In short:

A CORPORATION, LIKE AN ARMY, SHOULD BE DESIGNED SO THAT THE GREATEST POSSIBLE IMPACT OF THE WHOLE ORGA-NIZATION CAN BE QUICKLY AND SIMULTANEOUSLY CONCEN-TRATED ON THE MAIN ISSUE—THE OVERRIDING OBJECTIVE TO BE ATTAINED, THE BIG PROBLEM TO BE SOLVED, THE DECISIVE BATTLE TO BE WON.

Concentrated power wins battles in war and business, not numbers of soldiers or employees. "The cult of numbers is the supreme fallacy of modern warfare," wrote Liddell Hart. It's the weight of force at the key point that tells the tale. Even if the competitor is larger in total size but you can shift and maneuver more people and more skilled ones to a narrowly focused area more speedily, you can win.

But even the largest aggressive companies are bent on maximizing their concentration of force. IBM, for example, uses a system that it calls "resources balancing" to assign the right employees and the right numbers of them to carry out its strategic priorities. Merely getting the people to the right place is a prodigious and expensive undertaking. For example, in 1985, IBM moved more than seven thousand of its employees at an average cost of sixty thousand dollars each.

The saying goes, "Put your money where your mouth is." The principle of economy of force says, "Put your people (and other resources) where your concentration is." IBM's business is computers. Its IBM Credit Corporation (ICC) is a baby of a company—just seven years old. Yet it has assets of five billion dollars and racks up more than half a billion dollars a year in revenues. Its growth rate is more than 50 percent a year. It could be a giant if IBM wished it to be. But however successful it may be, it's minuscule compared with IBM's earnings of more than six billion dollars per year in computers, so IBM management purposely restrains its growth. It puts its resources where its concentration is.

Monolithic Dart & Kraft, the almost-ten-billion-dollar consumer-products and food corporation, broke itself into two companies when it realized it had two distinct concentrations—Kraft Incorporated would concentrate on foods, and the other, Premark, would assume all the other operations, from Tupperware to West Bend appliances.

Concentration won't happen unless you distribute your forces accordingly. ALL OTHER FACTORS BEING EQUAL, THE CORPORATION THAT ASSIGNS ITS EMPLOYEES TO GREATEST ADVANTAGE SHOULD WIN. Napoleon was the master of concentration. Even he said so. He won by outconcentrating his enemies because he was better at getting the most people to where they were most needed—

at the point of contact. Napoleon frequently confronted an enemy that outnumbered him. This didn't faze him, because he knew that *total* numbers didn't matter as long as he made certain that he skillfully shifted his manpower so that he outnumbered the enemy wherever he attacked.

For example, five separate battles were fought in the Montenotte campaign (1796). Overall, Napoleon's army was outnumbered by a substantial margin of sixty thousand to thirty-seven thousand. But except in one instance, wherever *a battle* was fought, Napoleon had made certain that he outnumbered the enemy. In one battle he had ten thousand troops on the field, the enemy four thousand; in another he had eight thousand, and they had one thousand; in a third battle he outnumbered them three to one (twelve thousand to four thousand); and in a fourth he held an advantage of twenty thousand to his enemy's twelve thousand. The fifth battle was even, six thousand apiece, but only because one of Napoleon's officers had failed to bring up his troops in time. Otherwise, Napoleon would have had ten thousand men there ready to fight.

How are corporations applying the principle of economy of force by distributing forces to produce the greatest possible concentration at one crucial point? Here are a few examples:

- Corning Glass Works was operating at 2 percent margin and didn't have any money to invest in its future until it sprang from its base in ceramics and glass into richer fields of heat-resistant ceramics and fiber optics. The latter is Corning's future, and to succeed in it requires a concentration in research, on which Corning expected to spend up to approximately one hundred million dollars last year.

- Heinz puts increases in income into its "superfund." It draws from the fund to allocate money to develop new, particularly promising products.

- Monsanto's biggest problem in the recent past has been the lack of new products. Thus, in 1986 it significantly increased its corporate R & D budget to more than half a billion dollars.

- The fierce competition among computer firms has resulted in a shift in emphasis from manufacturing to marketing. Recently Data General Corporation laid off nine hundred administrative and manufacturing people but added one hundred people to its sales force. Another computer firm cut almost one-quarter of its work force in 1986, then hired one hundred salespeople. Apollo Computer eliminated three hundred jobs in 1985, but increased its sales force by 40 percent. Hewlett-Packard didn't add sales staff in 1986 but increased the efficiency of those it already had by eliminating unnecessary paperwork.

The best mnemonic for this principle of economy of force is "YOU CAN NEVER BE TOO STRONG AT THE DECISIVE POINT."

Proportion Your Manpower to the Objective to Be Achieved

A war should only be undertaken with forces proportioned to the obstacles to be overcome.

Napoleon

While you can never be too strong at the decisive point, it's easy to be too strong elsewhere. After graduating from high school and needing money for college, I went to work in a small factory. One of the first things I noticed was that the four stock clerks were rarely busy for more than two hours a day, while the lone shipping clerk was always swamped and never able to catch up. At the end of the day there were always shipments left to be packed and sent off—and a customer somewhere waiting. That's just one tiny factory, but huge, sprawling corporations are discovering just how easy it is to be too strong in the wrong places. For example, at the moment, Madison Avenue is not the place to be. Corporate advertising budgets are down, and advertising firms are quickly seeking to be strong at other decisive points. Some are shifting toward offering public relations services; the Ogilvy Group is moving into marketing.

Proportions and Flexibility

The first sign of poor proportionment of manpower to objectives is idleness. If any part of your work force is idle or more idle than another, it's probably not where it's needed most. Another sign of a proportionment problem is an imbalance in the ratio of your "back" to "up" personnel. "Up" meaning people actually in corporate combat—producing and selling, for example; "back" meaning support of those "up." The ratio goes haywire fast. At the end of World War II, for every man up in combat, more than ten were back. In the late 1960s in Vietnam the seventy thousand men in combat units were being supported by six hundred thousand "back."

The common complaints about municipal repairs—two people in the truck while only one is outside working—is a well-known "up" versus "back" problem. In most companies, while the premium should be placed on what's up, the number that's back is the problem—too few people on the line, too many at headquarters complaining about them.

Having fewer staff "back" correlates with upping sales. In 1985 the management-consulting firm of A. T. Kearney, Incorporated completed a study that revealed that competitively superior corporations on the average had five hundred *fewer* headquarters staff per one billion dollars in sales compared with poorer competitors.

When Edward J. Noha began reshaping a faltering CNA insurance Corporation, he quickly reduced the number of officers from three hundred to fifty. It's disgraceful, he said, "but it took me several months to figure out how many people we had and where they were." Dana, one of the excellent companies cited in *In Search of Excellence*, solved its up-to-back inequity by reducing its corporate headquarters staff from five hundred in 1970 to about one hundred today.

To concentrate the full potency of an entire company on the main issue requires organizational flexibility. There are no two ways about it. In chapter 3 we saw that no plan should be expected to survive the first contact with reality, not even a highly detailed plan. Contact with actuality is tremendously forceful feedback, and it *always* calls for two things: quick on-the-spot adjustments in the field by opportunistic managers and planners racing back to the drawing boards to bring their handiwork into sync with what's really occurring. Because that's the case, no organization should remain exactly the way it was before contact. Your experience tells you that you can't possibly predict the form that a competitive opportunity or threat will take or exactly when either will appear. Your only reasonable recourse is to be ever vigilant as to possible threats and opportunities and to develop the capacity to shift money and manpower toward them more quickly than your competitors.

In her book *The Change Masters*, Rosabeth Moss Kanter asserts that the successful organizations of the near future will have more "surface" meeting the environment and will have numerous mechanisms for detecting emerging changes. Above all, these organizations will be flexible in bringing resources together quickly. Inflexible companies make impossible demands. They say to their environments, "We're structured the way we're structured. Whatever changes happen out there better fit into one of our department pigeonholes or we can't deal with them." Flexible companies know better. They're like the amoeba, with no fixed boundaries. They're fully prepared to adapt their structure—and readapt and readapt again. They say, "Whatever changes out there, we can quickly adjust."

THE IMPORTANCE OF STRUCTURE

Little is known about the life of French colonel Ardant du Picq except that he was killed in action at the age of thirty-eight while leading a regiment in battle on August 15, 1870, and that he had an insatiable hunger for facts. The facts du Picq was obsessed with were the basic realities of battle. What was battle actually like? What happened? Why exactly did one side win and the other lose? Du

Picq's words jump at you. His ideas are clear and vivid; they have the touch of reality to them.

Among the many questions du Picq puzzled over was why the Romans, who were not particularly brave people, who fought with weapons that were no better than their enemies', and who were usually outnumbered, regularly defeated larger armies of braver people like the Greeks and Teutons? He discovered the answer in two factors: the strict discipline of the Romans and their organization—the Roman legion. Maneuvering from their legions, the Romans were able to perform a consistency of battlefield maneuvers that couldn't be matched by the less organized foe. Du Picq realized what great commanders know and what corporations are learning anew each day: the side that's better organized tends to win. By better organized is meant organized correctly to implement its particular kind of strategy.

Even the most ancient battle of which a detailed account remains demonstrates that organization wins: Pharaoh Thutmose III's victory over the Syrians at Megiddo, Palestine, in 1479 B.C. Thutmose attacked in a concave formation, then outflanked, enveloped, and destroyed his enemy. The Assyrians, too, won because they were better organized than their adversaries. Like a modern corporation comprised of specialists all working together and coordinated, the Assyrians were able to orchestrate the division of labor of their spearmen, archers, and cavalry and charioteers into a battlefield presence of overwhelming strength and flexibility.

Every strategist—one working for GE or Quaker Oats or for the Romans or Assyrians—has had to wrestle with the problem of how to structure his personnel to match strategy, how to achieve economy of force along the lines of strategic concentration.

Layers

More than three decades ago (1954), Peter Drucker's book *The Practice of Management* suggested that no efficient organization required more than seven layers. The 1985 A. T. Kearney study mentioned above showed that the better competing corporations on the average did have seven layers, while the losers generally had more than half again as many layers, 11.1 layers on the average. The trend today is toward even fewer than seven corporate levels. Five appears to be the number that "downsizing" layer-reduction programs are leading to. Ford has seventeen; Toyota has five. Dana Corporation quadrupled its sales during a twelve-year period at the same time it was reducing its layers from fifteen to five. B. Charles Ames, chairman and CEO of Acme-Cleveland Corporation, says, "As a general rule, I think that whenever you have more than five layers in an organization, there is probably something wrong."

Genghis Khan achieved what few military commanders and corporate leaders of today could duplicate. He transformed a mass of unorganized bandits into the greatest army the world would see until four hundred years after his death. And he achieved the transformation primarily through his genius for organization grounded on quality, not quantity.

His armies were divided into tens:

- A division of ten thousand was called a *tumen.*

- Each division was comprised of ten regiments, called *minghans.*

- Each regiment contained ten squadrons (*jaguns*) of one hundred men.

- Each squadron was divided into ten troops of ten men called *arbans.* Army, *tumen, minghans, jaguns, arbans*—the terms may be different, but the organizational layering idea is the same: FIVE LAYERS.

The ten men in each troop elected their commander, and the ten commanders of troops elected the commander of the regiment. Each regiment and division commander was appointed by the khan.

Genghis Khan's invention of an army divisible by tens matched his strategy of mobility. It gained him flexibility through division of labor. Since World War II there has been a trend among all armed forces toward increased flexibility and mobility through advanced technology, improved communication, and perhaps most importantly, the development of adaptable organizational forms. Corporations, too, are pursuing the same trend toward malleable organizations.

In his landmark book *Strategy and Structure*, Alfred D. Chandler of Harvard demonstrated that the form organizational structures take in huge firms such as Sears, Du Pont, General Electric, and General Motors are responses to changes in the marketplace. As new product lines are developed, companies divide themselves into decentralized divisions. They change structure to accommodate strategy. In their famous study, described in *Organization and Environment*, Paul Lawrence and Jay Lorsch came to approximately the same conclusion as Chandler. Particularly in faster-changing dynamic environments, firms that succeed better than their competitors have developed superior mechanisms for adapting to economic and market conditions.

Efficient organization always involves three considerations:

1. Determining the strategic objective. Organization serves objective. Without objectives chaos reigns and any organizational form will do just as poorly as any other. As your objectives change, so should your organizational structure.

2. Setting up the structure.

3. Assigning resources.

No matter how large or small your company or unit is, it's worthwhile periodically to stop and ask yourself the following questions about organization.

- Is each part of the company (or the unit) really needed to achieve your "must" objectives?

- What activities does the organization do well? What does it do poorly?

- Does each part of the organization contribute to the whole without seriously duplicating the missions of other parts of the company?

- Where do the high costs lie? Where could you cut back?

- Is each part coordinated with the whole?

- Is each flexible enough to react to changes in the environment, even rapid changes? Even completely unforeseen changes?

- Is each making efficient use of personnel, money, material, facilities, and equipment?

- And the most essential question of all, "Why should you keep the organization the way it is?"

Seek Smallness Among Bigness

The lesson military leaders teach is conveyed in the maxim leading off this chapter: "It is not big armies that win battles; it is the good ones." Beyond some particular point, economies of scale are exceeded. Superior force has nothing to do with numerical superiority. One thousand men will defeat one, but one thousand won't necessarily defeat even one hundred. In chapter 3 we saw Darius, with perhaps one million men, losing to Alexander the Great with an estimated forty-seven thousand. And in chapter 4 we learned that in many of the decisive battles of history the victorious side had fewer troops and the loser had more. The economy-of-force principle explains why: the smaller winner had all his troops where they were needed; the larger loser had his too thinly dispersed.

Big corporations, like large armies, aren't necessarily as formidable as they appear. The problems of organizational unwieldiness, confusions among managers as to end objectives, poor coordination of effort between units, slowness of response to market changes, and the sometimes astronomical cost of maintaining size—these are all well-documented drawbacks of big corporations.

All social systems, including corporations, have an innate tendency to grow larger. Better companies resist the tendency toward size for size's sake. PRUNING—businesses and staff—is frequently the first order of business of the executive seeking improvements in economy of force. It's one of the first executive actions we saw in

Waging Business Warfare when Gideon fired ninety-seven hundred to run a bare-bones, lean-and-mean army of three hundred top-notch fighters.

Sherwin-Williams in paint and Allied Products and Caterpillar in farm equipment are just a few examples of companies that pruned to achieve the slimmer look. Gulf & Western is a ponderous conglomerate that's being restructured for the same reason. ITT has sold eighty-five of its subsidiaries. GE, R. J. Reynolds, Goodrich, RCA, Texaco, Beatrice, Continental Group, Dow Chemical, Du Pont, United Technologies, and Mobil are among the largest corporations that have each sold off substantial business interests in the last few years.

In 1985, Atlantic Richfield, "barely breaking even," pruned more than one billion dollars in unproductive assets, including all of its ARCO gas stations east of the Mississippi. Greyhound Corporation's chairman John Teets has sold more than three hundred million dollars of the conglomerate's assets.

Genghis Khan invented the Asian version of the divisible army in the thirteenth century; the French devised the European rendition in the 1760s in the form of the division. The reason was the same as why corporations split off into divisions. The divisional structure put an end to the single huge mass forming a solid front on the battlefield. Under the command of a general and self-contained and strong enough to take on the enemy until reinforcements arrived, the detachable division opened new flexibilities and mobile possibilities that were impossible under the older form of "one mass" organization.

When Napoleon became emperor, his first order of business was changing the structure of his armies along divisional lines. Napoleon didn't invent the divisional structure, but he put it to better use than anyone before him had thought possible. French commanders had been experimenting with divisional arrangements for almost three-quarters of a century before Napoleon showed how it really should be done by making each division a semi-independent, self-contained instrument of war. Starting with the campaign of 1800, he organized corps of two or three divisions, which he placed under the command of lieutenant generals and formed into the wings, center, and reserve of his army. The divisional configuration, Napoleon's secret weapon, made his strategy possible, just as Sears's facilitates its strategy. Napoleon's divisions meant he could move his army in separate columns with separate lines of advance rather than in one huge, slow-moving, moiling mass. Divisions could move faster, disperse more rapidly, and converge more quickly at the point of concentration.

Commentators on corporate strategy point out that "surprisingly, organization design tends to lag behind strategy formulation. It is employed only sporadically as a tool for promoting strategic change, and too often existing organization actually holds back the execution of new strategy." Napoleon saw the dangers inherent in strategy without a matching structure and refined his divisions.

Large corporations are developing new uses of the flexible divisional form, the basic organizational design that best facilitates the implementation of corporate strategy. Many are breaking themselves down into smaller, semiautonomous divisions that match their strategy.

For example, Johnson & Johnson had 80 divisions a decade ago and now has 150. Acme-Cleveland Corporation has created twenty profit centers from an original five and is attempting to create more. In early 1986 huge Exxon, the world's largest oil corporation, began a massive program to reduce managerial levels, increase the speed of making decisions, and shrink the size of its workforce.

WARRIORS AND SOLDIERS

The difference between a warrior and a soldier is numbers. A warrior is a loner, a rugged individualist. He is a generalist, not a specialist. He is like the person in a one-person operation or an entrepreneur working alone on his deals. He needs to organize his time and priorities, but he's not an organization.

But the moment two corporate warriors agree to work together and to subordinate their individual actions to a larger whole—if only "You're better with people, so you handle the selling; I'll take care of operations"—that moment an organization is born and the "warriors" are now "soldiers." If all goes well the company hires other soldiers, then more and more. And then they develop division of labor, hiring specialized soldiers—some infantrymen, artillery experts, logistical support specialists—planners, sales people, marketing specialists, and managers, etc. Each soldier has a different specialization, and each is given a different span of surveillance and area of responsibility. And soon you have GM or Exxon.

No one can deny that specialization and division of labor have reaped marvelous rewards for corporations, and for armies. Total lack of specialization is as much a sign of a primitive level of development as is the lack of specialized machines. But jobs that are too narrowly defined reduce productivity.

The problem is that by specializing too much, corporations are overlooking the fact that the warrior is a more complete fighter than the soldier. He can do more; he's versatile. He cannot match up to the soldier's ability in the soldier's particular specialization, but in virtually every other skill of war he's far superior.

Underlying the high-specialization school of thought is the belief that the average person has a special capacity for a particular job. This completely overlooks the virtually unlimited ability of the average person to adapt to jobs, to learn and to master tasks that are even quite remote from his or her previous experience, and to perform admirably in more than one field of activity. A popular misconception challenged by the research study of Benjamin S. Bloom on the development of talent is that the talented person is extremely competent in only one area. The study indicated that this wasn't true at all. Generally the talented person is gifted in a number of areas.

I saw the same thing when I worked as a consultant with a research team for the University of Michigan's Institute of Labor and

Industrial Relations. We found that the more productive of the organizations we were studying encouraged "role overlap" where the lines of demarcation between one function and another were blurred and functions overlapped. For example, although Sally was a customer-relations specialist and Bill a salesman, Sally would sometimes sell, and Bill would handle customer concerns. Both, in short, possessed some of the versatility of warriors. Less productive organizations operated differently. They attempted to make as certain as possible that staff remained within the confines of their specific function.

The Japanese are famous for their ability to react quickly to changes in the corporate environment, such as advances in production technology, crises, and competitors' moves. And they are equally famous for their version of role overlap. They tend to use personnel as generalists (warriors) rather than strictly specialists (soldiers). Procter & Gamble makes a profit through role overlap. When its Charmin Division salespeople make a call, they not only sell bathroom tissue but other P&G paper products. The company saves probably 2–4 percent of sales through doing so. Employing "warriors" rather than "soldiers" very directly affects the principle of economy of force to serve your concentration. You'll be better prepared to shift your personnel to the point of main effort if those people can perform more than one job. The winning side concentrates more firepower than the losing side. Fire is everything. So, in the military, even cooks are trained to fire rifles. They might never be called on to shoot their weapons, but if they are, they can.

Using Reserves

... conserve a decisive mass for the critical moment.
Clausewitz

On August 6, 1944, two months to the day after the Normandy invasion, the total number of American reinforcements ready for combat to replace any man killed on that day was one. One man. Many corporations find themselves in the same boat.

The purpose of reserves is to rush to the part of the line where the action is. Ferdinand Foch called the reserve a carefully maintained club to carry out the one act of battle from which a result is expected—the decisive attack. In 422 B.C. a Spartan observed that reinforcements are *always* more formidable to an enemy than the troops with which he's already engaged.

Military commanders generally select reserves before battle and hold them back until the critical moment. Corporations can do the same thing by designating certain units and employees as reserves "in case"—in case your offensive is thrown back and in case you're forced to hold off a competitor's assault or any shift in business conditions.

In the 1970s Hitachi needed an engineering reserve force in its information-based businesses. It simply redeployed engineers from one part of its diversified manufacturing enterprise. In 1986 Matsushita Electric Industrial Company experimented in human retooling, too. To support its shift from its traditional electronics and home-appliance business to its future business of semiconductors, plant automation, and office equipment, Matsushita began transforming forty middle-aged engineers into industrial salesmen. Contracting work out and making use of temporary workers through employee leasing are also examples of making use of reserves. Hewlett-Packard employs workers on a contract basis during busy periods—engineers, programmers, and technical workers; John Hancock contracts workers for rush jobs. It's obvious that the use of temporary reserves by corporations is on the increase. In total, more than 700,000 temporary jobs are filled every workday. In 1975, the figure was 186,000.

DEVELOPING PEOPLE

*One should attend to his warriors as he would to his
own thirst.*

Samurai Takeda Nobushige

The war principle of ECONOMY OF FORCE involves distributing your forces to the greatest advantage. And this principle, in turn, involves managing people. Whatever else his or her corporate role involves, the manager of a company unit, like a military commander, must create a group of human beings who are ready to win now! A company's strategy must be accompanied by a matching organizational structure, and that combination of strategy/structure must be supported by putting the appropriate people in charge and giving them people who are equipped to carry out the strategy. All aspects of human-resource development (HRD) figure into the economy-of-force principle, including management training and development.

Traditional training for managers is intended to enhance management potential. Courses usually include conference leadership, work simplification, speed reading and comprehension, quality control, human relations, report and letter writing, etc., and the old standard time management, etc. A number of corporations are questioning this shopping-list approach and are seeking to tie executive development directly to the company's strategy. General Foods, Xerox, Federated Department Stores, and Motorola have shifted away from the traditional emphasis on training designed to improve management potential to course work specifically to implement the company's strategy and achieve results.

In 1906, philosopher William James said, "The world . . . is only beginning to see that the wealth of a nation consists more than anything else in the number of superior men that it harbors." The same could be correctly said today of corporations. All corporations and armies compete for talent, then compete with it. Clausewitz related human-resources development directly to the success of a strategy when, in *On War*, he pointed out that the principal way to keep strategic action moving is to develop additional manpower: ". . . there is in strategy a renewal of effort and a persistent action, [which] as a chief means toward ultimate success, is more particularly not to be overlooked, it is the *continual development of new forces.*"

Understanding Human Nature Is a Bitch

One of your jobs as a manager is to make full use of the capabilities of the people assigned to you. It's no secret that it's foolhardy to ask people to do what they don't have the ability to do or to do for any length of time something that doesn't suit their personality. Physiologically and anatomically humans are similar. It's in psychological and spiritual areas that individual differences are glaring.

One of the most revealing stories of the differences in human nature comes from the account of twenty-three American military men and two Eskimos who, in August 1881, established an isolated camp on Lady Franklin Bay in an attempt to live closer to the North Pole than civilized man had lived before.

Under the command of 1st Lt. A. W. Greeley, the expedition was expected to be taken out by rescue ship after one or two years. But the harsh winters coated the waters with thick ice that led to the failure of two rescue attempts. The small band of men would have to hold out or die. During all of the first year and well into the second the men remained in reasonably decent spirits, although slipping at times into sleeplessness, irritability, melancholy, and a sense of monotony. Discipline was kept strict, rules of subordination and delegation were maintained, and classes were organized to keep the men's minds stimulated. There was sufficient food, and a sense of camaraderie prevailed. Lieutenant Greeley created the first rift in morale when he unwisely forebade the men from going more than five hundred yards from camp without permission. A slight tear in the fabric of group morale and cohesion appeared, but it wasn't serious.

August of the second year passed with no sign of the relief ship. Greeley decided to wait no longer but to move to bases that supposedly held food supplies. One disaster after another befell the Greeley expedition from that point on. When it was finally rescued on June 22, 1884, only seven men remained alive, and they were so weak from lack of food that they hadn't the strength or will to raise the tent that had blown down on them. They had resigned themselves to die. Two of the men passed away shortly after being rescued, leaving only five.

Of the twenty who perished, most had died of slow starvation, but some had been shot, and others had died attempting to save the lives of the others. A diary kept by one of the men describes the roles played by the various members of the group.

Greeley was a strict disciplinarian who commanded the bodies of the men but not their spirits or loyalty. That any men survived at all was not due to Greeley but to Sergeant David L. Brainard, who lived to become a brigadier general. Brainard supported Greeley to the end but possessed a personal warmth and human understanding that Greeley lacked. Brainard softened blows, extended a helping hand, and rallied the men when they were down. He was not physically the strongest or necessarily the most courageous, but he was compassionate and had an unbreakable spirit.

The camp physician, highly trained and well educated, had been chosen for the mission because he got along well with the men. But under the ordeal he became the most self-centered and self-serving of all. Had he not starved to death, Greeley would probably have had him executed.

Jens, an ignorant Eskimo, was the most decent of all. He died trying to carry out a rescue mission. Some of the others had no pity for the weaker men and lost all sense of duty and discipline. They abused, cheated and robbed, and even stole weapons, intending to murder the others. After repeated violations they were ordered shot.

But however powerful their urge to preserve themselves and simply to live, if even for one more day, the majority maintained a sense of honor and dignity. One, Private Schneider, was so concerned with his integrity that just before dying of starvation, he left a note reading, "Although I stand accused of doing dishonest things here lately, I herewith, as a dying man, can say that the only dishonest thing I ever did was to eat my sealskin boots and the part of my pants."

Another, Private Fredericks, was brutal and vicious to others during the first year but actually became a better man as conditions grew worse. When the others were in a state of physical collapse or in a coma, Fredericks became a pillar of spiritual strength, aiding and comforting those comrades who had fallen. Lieutenant Lockwood was strong and resolute in the early days but totally beaten toward the end. Sergeant Gardner often gave his scraps of food away to whoever he saw needed them more. Private Whisler died begging the others to forgive him for pilfering a few slices of bacon. Private Bender was alternately a hero and a scoundrel.

The bravest and noblest of all was Sergeant Rice. His spirit never weakened, his conduct was incorruptible and his will indomitable. He performed one act of selfless heroism after another and froze to death on one last, hopeless mission through a storm to find a cache of seal meat for his friends. Private Fredericks, who was with him when he died, described his last moment with Rice: "Out of the sense of duty I owed my dead comrade, I stooped and kissed the remains and left them there for the wild winds of the Arctic to sweep over."

There you have it—an American work force. Admittedly, it was one forced to endure the cruelest of conditions, circumstances that virtually no corporate unit will ever face. But nevertheless it was a team of real people who revealed qualities that members of any company's work force demonstrate as they work together over the years. There is Greeley, the theory X authoritarian manager, and his theory Y right-hand man Brainard. On the one hand, there are generous, helpful, fair supervisors and dutiful, hardworking, and high-spirited men, some of whom are respected, even loved, by their fellow workers and some who come through under pressure time and again, achieving a very real kind of greatness. On the other hand, there are the shirkers and rule breakers, the sappers of morale who are totally devoid of any sense of loyalty to the group—the "me first, the hell with you and the company" type. And there are the bad workers who turn good; and those who start out well and turn sour; and others who are sometimes fine and at other times bad apples.

The need to understand individual differences in ability and performance predates civilization and harkens back to the earliest group enterprises. Surely when a band of cavemen ventured out to slay a beast for dinner it was recognized that Opu handled the club more skillfully than Delf, who was, however, a better tracker than Sempf, the best planner of the tribe. The Chaldeans made deployment decisions on the basis of accidents of a birth—a person's horoscope—and long into the nineteenth century personality assessments were often phrenological. Phrenology was a supposedly objective assessment of an individual's personality, temperament, and motivation derived from measuring the bumps on one's cranium. Mandarin China based selection and promotional decisions primarily on the basis of the person's writing ability. Vegetius advised that the person selecting new levies should look for recruits with lively eyes.

Generally, people like to consider themselves good judges of character even without feeling the other person's head for bumps. Most managers with any reasonable experience and savvy at all can predict fairly accurately which corporate unit will produce more than another and which employees are the best producers and which are the poorest. You know, for example, that Mary will put out half again as much work as Bill day in, day out. And you realize that if the new project calls for creative ideas, Harry is the guy to look to, but if it demands someone who gets along well with others, Charlie is the person you best call on.

If you know your people and can accurately assess their strengths, you can match them with assignments so expertly that you surprise, even dazzle, other people. For example, when the British Cabinet was considering sending a force to Burma to take Rangoon, the Duke of Wellington was asked who would be the best general to command. "Send Lord Combermere," he said.

"But we have always understood that Your Grace thought Lord Combermere a fool," replied a government official.

"So he is a fool," replied Wellington. "A damned fool, but he can take Rangoon."

But there is a great deal of information supporting the notion that understanding human nature is a real bitch and that there is no way of telling precisely what people will do, particularly when the chips are down. For example, the physician was specially selected for the Greeley expedition because of his ability to get along with the men. Yet as events unfolded, he turned out to be devoid of any sense of compassion for others whatsoever. During the second battle of Le Cateau in 1918, commanders of the British infantry judged that the troops were too exhausted to fight again. But as the cavalry passed through to take up the attack, the infantrymen rose to their feet, cheered, and advanced, something their officers had not expected.

You want to be able to predict how a person will perform in the future. What indicators do you use? The first thing you do is probably look to what he's done in the past—only to find that people can screw up the works by changing. As a prince, Frederick the Great devoted himself to poetry and gardening. When he ascended the throne of Prussia in 1740, he surprised everyone when he ceased his "idle pastimes" and suddenly transformed himself into one of the world's greatest warriors bent only on creating a powerful Prussia. Shakespeare's Henry V is a play reminding us of the thin ice we're walking on when we try to judge a man by his past. A bust as a prince, Henry became a fine king.

Research on combat in World War II revealed that in company after company the soldiers who during training were lazy, disorderly, and unruly *consistently* became the best and most disciplined fighters on the battlefield. When the engagement had ended, they almost invariably reverted to their undisciplined ways once again. Great at fighting, they were terrible at soldiering. This always makes me think of my own experience training thousands of salespeople. Time and again the best producers turn out to be the very ones whom the managers tell me give them headaches, are a little less manageable, less prompt with reports, who stretch the rules, etc.

In business we would expect that gripers would be among the lower performers. Yet research on a large insurance company and an appliance factory reveals that the best, most productive workers were considerably more critical of company policy than less productive workers.

How a person has performed in educational settings is commonly used as a basis of forecasting managerial performance. Should it be? The voices of those who answer no become more numerous by the day. For example, in a *Harvard Business Review* article entitled "Myth of the Well-Educated Manager," J. Sterling Livingston tries to explain why so many MBA holders experience "arrested career progress," while less well educated people often get to the top in management. He asserts that academic-type achievement is not a valid predictor of managerial potential—not the number of degrees a person holds and not his grade-point average or the management programs he has attended. He concludes that the people who reach the heights of corporate management have developed skills that aren't taught in courses and may even be quite difficult for highly educated people to learn.

Educational achievement is one type of credential being called into question as a tool for judging who will perform best and most consistently. Other traditional credentialing systems are undergoing the same type of reevaluation. Who creates and innovates? According to Edwin H. Land, the inventor of Polaroid, it isn't necessarily the person with the best educational background (he himself quit school before graduating) or the highest IQ. Discoveries are made, Land said, "by some individual who has freed himself from a way of thinking that is held by friends and associates who may be more intelligent, better educated, better disciplined, but who have not mastered the art of the fresh, clean look at the old, old knowledge."

If it's not necessarily a person's horoscope, head bumps, liveliness of eye, prowess in training, degree of griping, educational achievement, intelligence, etc., that tells us how well the person will come through in the future, what is it? What can be used as a basis for predicting future performance? And what are managers really hoping to instill in their work force?

If we look at military history, the single most pertinent factor is a constellation of subjective human qualities that varies considerably from person to person but can be increased by good leadership—"the fire and movement habit."

Find People with the Fire and Movement Habit and Permit Them to Advance and Shoot

In four months' time Bell & Howell's Documail unit produced a high-performance sorter for the U.S. Postal Service (USPS) that would have taken most companies twelve months to design, engineer, and build. The company won another USPS contract by having its employees at the post office twenty-one hours a day, five and a half days a week for eleven weeks. Bell & Howell is the only USPS supplier that ever delivered machines on time at specifications. Nothing fancy, Documail won by working harder than its competitors on both sides of the Atlantic. The unit didn't forget that human sweat and energy spell competitive victories.

In warfare, forward movement and fire win battles. In the military, the person who has the fire habit is always looking for ground ahead of him that will increase the effectiveness of his fire. Good troops always want to get into the action. One of the main forms of competition within the competition of business is over which side has the superior movers and firers—which has the higher producers. High productivity is an offensive weapon. It is never, never equal on both sides.

Moving forward and firing are the equivalents of the principles of offensive action (chapter 5) and concentration of force (chapter 4), but on the level of the individual. Companies, like armies, will go only as far on the offensive as their personnel will take them, and not one inch farther. And no corporation has a ghost of a chance of breaking through at any point of competitive concentration unless its managers and other workers "fire" through concerted human effort. The same is true in warfare, yet when an infantry unit engages the enemy, 85 percent of all the men will not fire their weapons at all or will fire only when an officer is standing over them. Only 15 percent of the men in any unit in any army at any time or in any place have the fire habit. Air forces are even worse off. Half of the enemy planes are shot down by a mere 5 percent of the fighter pilots. To shoot down the other half takes 95 percent of the pilots. In business warfare the situation isn't so terribly different. In company after company, 80 percent of the orders are turned in by 20 percent of the sales force. In some fields, 90 percent of the salespeople don't even attempt to close—don't even try to fire away and ask for the order. In Eskimo warfare, if one side wished to take a break from fighting, it simply hoisted a fur coat up on a stick as a signal. Reports on diminishing corporate productivity tell us that the ratio of nonfirers to firers is high in business, too, and suggest that a lot of fur coats are being hoisted on sticks.

All you need do is look at any group of people doing the same kind of work. The best of all will be found to be producing between two and ten times as much as the poorest and slowest. I think that's a startling statistic, yet when I cite this figure to managers, they often say I'm understating the case. They insist that some of their staff outperform others by a far greater margin! This wide differential of productivity is found in all occupations. Research on the pace with which work is performed has discovered that no "average" pace can be established that suits most humans. Work pace is as individual as a fingerprint. Some of the differences in a person's orientation to aggressive, forward-moving action is also no doubt innate. There are fox-terrier managers and workers and beagle managers and workers. Among breeds of dogs, fox terriers are better fighters than beagles, though beagles are bigger. You can take a litter of beagles and a litter of fox terriers born on the same day, mix them up, putting half the beagles with half the fox terriers, and raise them all the same way. In less than four months the smallest beagle will be larger than the biggest fox terrier. Yet pair up a beagle with a terrier in competition for one bone and the smaller terrier will almost invariably get the bone. Exactly the same age as the beagle and reared precisely the same, the terrier will turn out to be smaller but more aggressive.

The most effective managers are probably the ones who, realizing that some beagle workers want to be beagles, will find beagle jobs for them and will give the fox terriers tasks appropriate to them. Meyer Rothschild, the founder of the House of Rothschild, was that kind of manager. He had four sons. Each was very different in ability and personality. Rather than dwelling on their shortcomings, he decided to assign them by matching their strengths with the necessities of the assignment.

Nathan was the ablest and most daring, tough and aggressive. The best fit for his personality was London, where the wheeling and dealings were done by tough, hard-nosed businessmen. James thrived on intrigue and for years had been the political strategist of the family. The best place for him was Paris, where business and politics blended in a web of intrigue. The third son, Solomon, was assigned to Vienna, where his courtly manners and personal charm would be accepted by the aristocrats. Amschel was assigned to Frankfurt, the least demanding of the financial centers, where his hardworking but colorless style would be put to best advantage.

Aggressiveness translates into motivation, and there is no factor that plays a more important role in explaining the Grand Canyon-wide differences in the fire habit as motivation. When all other elements of human difference are held constant, the effect of motivation is decisive. If two groups whose members are equal in intelligence are given a problem to solve, the more highly motivated will do a better job—a much better job.

Vegetius wrote, "The expense of keeping up good or bad troops is the same." In fact, the cost of maintaining bad workers is higher. One manufacturing company estimates that 21 percent of its labor cost goes into paying to correct mistakes made in production.

Some people will always find a reason for staying put and not budging. If a Choctaw brave wanted to avoid battle, all he had to do was cite an omen such as a bad dream. Whole war parties returned to their home village because of an unfavorable dream. In business warfare the equivalent of a Choctaw bad dream is citing anything that throws a scare into people. Convincing decision makers that the idea will cost way too much, will never fly with the CEO, will cause labor tensions, etc., are modern corporate versions of a not-so-brave brave's unpleasant dream.

Below are capsule lessons on moving and firing from the military masters:

- Put the best mover and firer in the lead and have the rest follow.

- Instill a healthy contempt in your work force. In numerous cultures there existed a death contempt. American Indians called the braves who were contemptuous of dying "no flight men." They went into battle to fight, not to run away. Groups of such men were called "no retreat societies." In business warfare these are men and women who feel so strongly about achieving the objective, doing the job, and carrying out the project that they are contemptuous of the consequences. They are totally committed. They don't give up; they don't retreat.

- Make firing a habit. Habits are developed by repetition, so that the action becomes automatic. If your people are not aggressive movers and firers, it's because they've gotten into the reverse habit. The idea is to replace those habits with good ones by repetition. This can best be done by training. It's axiomatic in warfare that a soldier should never encounter in battle what he has not seen first in training. "The more you sweat in peace, the less you bleed in war." Training is another form of competition within the competition of business. It's not mere coincidence that Japanese firms spend between five and seven times the amount of money on staff training as their American counterparts. The Prussian army of Frederick the Great gained substantial advantage over its adversaries because it was better trained. The average Prussian soldier was trained to load and shoot his musket twice as rapidly as his counterpart on the other side.

- Maintain a policy of promotion from within. Such a policy has been shown time and again to lead to high morale and productivity.

- Inspect along the lines. See for yourself if people are moving and firing, even if it means spending time with each person and directing his or her fire through feedback. If every worker was talked to in a positive, supportive, and reinforcing way about the importance of the job and his/her contribution—even for a few minutes every day—productivity would increase.

- Eliminate the causes of inaction. The U.S. Department of the Army has identified five reasons why soldiers fail to fire their weapons. Each has a parallel in business warfare.

1. Lack of confidence in weapons, or skill in their use. Training is the answer.

2. Hoarding of ammunition. The business equivalent is hoarding time—putting off for tomorrow what could be done today. Instill the idea that one today is worth five tomorrows.

3. Lack of the will to fight. Corporate staff can easily become lethargic when they feel too secure. Imbue your people with the fact that their effort translates directly into profits, that only profits bring security, and that the competition is after the same dollar your company's after.

4. Fear of provoking the enemy into retaliation; the desire to keep the front quiet. Communicate clearly the fact that the side

that wants to keep things quiet usually loses; the one that releases the first crackle of offensive gunfire ordinarily wins.

5. Apprehension about giving away one's position and drawing
   fire. Most often, the fire that a person fears in business is from
   his or her boss or fellow workers. Many actions that should
   be taken aren't simply because it would mean making waves.
   The fear of drawing fire in business war is fear of drawing
   someone's ire. Johnson Wax (S. C. Johnson & Son Incorporated) had a policy of lifetime employment long before the
   Japanese made such policies popular. It conducted a study to
   uncover blocks to innovation. One was found to be fear of
   someone's ire. Said Samuel Johnson, the company's CEO,
   "People have been hiding in the trenches with their new ideas
   —for fear of disapproval, fear of failure."

- Choose performance over position. A rigidly rank conscious
  army or corporation eventually falters. Truly responsible leaders select the best available person for the job, not necessarily
  the individual everyone believes is in line for it. One of the
  principal reasons Maj. Gen. George Meade was victorious at
  Gettysburg was because he had the guts to put General Hancock in command of three corps over more senior but less
  capable generals.

- Restrain less; unfetter more. Psychologists have demonstrated
  that if people believe they are in control of things, they will
  persist longer at tasks and perform better than if they believe
  control rests with someone else.

- Emphasize volunteerism. Militarily, volunteers tend to be good
  fighters. The same is true in corporations. People who volun-
  teer to do a job perform better and work more persistently to
  solve problems they encounter.

- Make them feel important. The ordinary Roman soldier believed he personally was important, and he fought accordingly.
  To arouse a person's commitment and use of his full range of
  skills give him a problem to solve that's larger than his narrow
  span of authority.

- Challenge them. Asked how Procter & Gamble manages to retain so many talented people, CEO John Smale said, "Well, first of all, it's got to be done on the basis of people enjoying being challenged by what they're doing. Talented people by their very nature need to be challenged. . . ."

- And challenge them early in their careers. One research study, for example, demonstrated that engineers whose initial job experiences involved challenging technical work tended to contribute to knowledge quickly and to maintain patterns of high competence and solid performance throughout their careers.

Make Ordinary People Extraordinary

The courage of a soldier is heightened by his knowledge of his profession, and he only wants an opportunity to execute what he is convinced he has been perfectly taught.

Vegetius

Among all the insights offered by the military masters none stands out as clearly as that even the most "ordinary" troops can become extraordinary fighters and that whether an army becomes extraordinary or not depends principally on what the leaders do, not what the soldiers do. Konosuke Matsushita is founder of the powerful Matsushita electrical empire, makers of Panasonic, Quasar, Technics, and National brands. He has said that his role was to help ordinary people become extraordinary. Matsushita borrowed those very words from a Japanese samurai general who lived three hundred years before him, Tokugawa Ieyasu.

That ordinary people really *can* become extraordinary is indisputable. The book *Developing Talent in Young People* is based on a five-year study of 120 adults who became high achievers in various fields before the age of thirty-five. The chief investigator, Benjamin S. Bloom, says that when he and his researchers began the study, they theorized that the high achievers would be found to have possessed "great natural gifts," to have been born more favored than those who were not as successful later in their lives. But the popular "natural gifts" theory didn't hold up during the study. Much more important than nature was nurture—the long and intense process of encouragement, support, education, and training. No matter what the characteristics of the people, without that type of nurturing process they would not have achieved the highest levels of capability in their fields.

Maintaining high standards of performance has also been shown to lead to higher levels of performance than maintaining lower standards. Set the goals higher rather than lower, let your people know what you expect, keep them informed of how well they are doing, and suggest how they can improve their performance even more.

Another significant body of information on how to make ordinary people extraordinary has to do with the so-called self-fulfilling prophecy, or Pygmalion effect. In essence, it proves that people tend to achieve close to the level of the expectations held for them by significant authority figures. In "battles of any kind," says du Picq, "two moral forces, even more than two material forces, are in conflict." He will win who has the resolution to advance. But no one will willingly advance if he doesn't see any chance of winning. Man shrinks back from any effort in which he doesn't foresee success, du Picq adds. The people who advance move forward because they expect to win. The Pygmalion effect, based on hard research, reveals that a person's expectation of winning is often the result of a manager or supervisor communicating the expectation of winning.

Often even those who are familiar with the power of expectations limit its application. They usually believe it pertains to a manager's expectations of an individual employee or a small work unit at most. In fact, one leader's expectations of a huge body of people—an army or a company, for example—can transform that body into a force directed at achieving the leader's expectations of them.

CREATE FIRE TEAMS

- "It's essential that we take more risks," said Eastman Kodak's chairman Colby H. Chandler. Then he split Kodak's business into seventeen separate profit-center units. As part of Kodak's "entrepreneurial" approach, each unit is headed by a young executive with virtual total authority over the unit. The objective of each is to exploit opportunities for the new Kodak as it ventures away from its core photo business, particularly into electronics.

- Procter & Gamble's CEO, John Smale, created "business teams," *ad hoc* groups of its people from all company levels and departments. Teams solve problems, develop product ideas, and prepare plans.

- Ford's fast-selling Taurus car, the "car of the year" in 1986, was developed as a result of Ford management dispensing with traditional organizational structure and creating "Team Taurus." Instead of product planners, designers, engineers, and the manufacturing employees working on the project in the traditional sequential basis—one finishing his job before the next begins—Team Taurus brought representatives of all the specialty units together.

- Shenandoah Life Insurance found that it took twenty-seven workdays to process an application to convert a policy and that thirty-two clerks handled the paperwork. After grouping the clerks into teams of five to seven members to handle all steps, the process was reduced to just two days, and complaints about service all but disappeared. Now Shenandoah is handling 50 percent more applications than before.

These are just a few examples that represent the many more that could be used to illustrate who wins battles and competitions. Studies on the use of "self-managing," semiautonomous teams reveal that the team approach is 30–50 percent more productive than traditional organizational forms. Why? However much technology changes things, says du Picq, the answer to which side wins and which loses will always be found in the one thing that never changes—the heart of man. The human heart is the main point of reference for all the truths of winning, not strategy or tactics.

The lesson of the military masters and the corporate leaders with a genius for winning competitive battles is the same as to what wins. Small groups do—not a corporation but the tens or hundreds of groups of people working together to manufacture a product or solve a problem—the project team, the squad. And why do employees rally together and put out effort? Because they know and trust each other and because they don't wish to fall in the estimation of the person working next to them. "Four brave men who do not know each other will not dare to attack a lion. Four less brave, but knowing each other well, sure of their reliability and consequently of mutual aid, will attack resolutely," says du Picq. In a nutshell, that's the science of organizing workers.

NOTES

CHAPTER ONE

The Art of Business Warfare

PAGE

3   For, although they were richly rewarded: Sun Tzu, *The Art of War*, trans. Samuel B. Griffith (New York: Oxford University Press, 1971), pp. 24–25.

4   "the soldier, high or low": Karl von Clausewitz, *On War*, trans. Col. J. J. Graham, ed. Anatol Rapoport, following text of New and Revised Edition, ed. F. N. Maude (Baltimore: Penguin, 1968), p. 168.

9   "for it is the first,": Harry Holbert Turney-High, *Primitive War* (Columbia, S.C.: University of South Carolina Press, 1971), p. xv.

14   "Competitiveness means taking action": "What Welch Has Wrought at GE," *Fortune*, July 7, 1986, p. 45.

17   Motorola's position improved: Chicago *Sun-Times*, Aug. 1, 1985.

19   "Success in war is obtained": attributed to Francesco Guicciacardini, *Storia d'Italia*, 1564.

20   "Bigness is not a sign of strength": "Splitting Up," *Business Week*, July 1, 1985, p. 53.

CHAPTER TWO

Leading

24   Similarly, Chester Barnard: Chester I. Barnard, *The Functions of the Executive* (Cambridge, Mass.: Harvard University Press, 1968), p. 198.

24   "It is not the Prussian army": J. F. C. Fuller, *A Military History*

of the Western World, Vol. 2. (New York: Funk & Wagnalls, 1955), p. 194.

25   "I was looking for a man": "Jerry Tsai: Comeback Kid," *Business Week*, Aug. 18, 1986, p. 73.

25   "Most corporate executives I've met": *Ibid.*, p. 75.

26   British field marshal Wavell: Alfred H. Burne, *Art of War on Land* (Harrisburg, Pa.: Stackpole, 1966), p. 8.

27   "Even though these companies": Thomas J. Peters and Robert H. Waterman, Jr., *In Search of Excellence: Lessons from America's Best-Run Companies* (New York: Harper & Row, 1982), p. 13.

28   "If you put Jack": "Not Just Another Takeover—or Is It?" *Business Week*, Dec. 30, 1985, p. 49.

29   "We need someone who can give us": "A Man with a Screwdriver Operates on AT&T," *Fortune*, June 23, 1986, p. 122.

30   Hamilcar, the Carthaginian: Edward S. Creasy, *Fifteen Decisive Battles of the World* (Harrisburg, Pa.: Stackpole, 1960), p. 94.

33   Though it was a small company: Mariann Jelinek, *Institutionalizing Innovation: A Study of Organizational Learning Systems* (New York: Praeger, 1979), p. 78.

34   A study by R. Joseph Monsen: R. Joseph Monsen, "Ownership and Management—the Effect of Separation on Performance," *Business Horizons*, August, 1969, pp. 45–52.

37   "His victories teach": Baron de Jomini, *The Art of War*, trans. G. H. Mendell and W. P. Craighill (Philadelphia: J. B. Lippincott, 1862), p. 20.

37   "Overoptimism stems from a failure": Thomas L. Berg, *Mismarketing: Case Histories of Marketing Misfires* (Garden City, N.Y.: Doubleday, 1971), p. 8.

38   Extensive research has been done: The reader is referred to the works of David C. McClelland on the achievement motive, including *The Achieving Society* (New York: The Free Press, 1967).

38   Peter Engel: Peter H. Engel, *The Over-Achievers* (New York: The Dial Press, 1976).

39   Edwin E. Ghiselli's studies: Edwin E. Ghiselli, "Managerial Talent"
        in *The Discovery of Talent*, ed. Dale Wolfe (Cambridge, Mass.:
        Harvard University Press, 1969), p. 236.

40   The discount brokerage company: "Charles Schwab vs. Les Quick:
        Discount Brokering's Main Event," *Business Week*, May 12, 1986,
        pp. 80–83.

41   In his studies of emotional problems: Michael Maccoby, *The
        Gamesman* (New York: Bantam, 1976), p. 200.

41   When a young IBM executive: Warren Bennis and Burt Nanus,
        Leaders: The Strategies for Taking Charge (New York: Harper &
        Row), p. 76.

43   He describes a number of types: James MacGregor Burns, *Leader-
        ship* (New York: Harper & Row, 1978), p. 244.

43   For example, one early morning: "Texas Instruments Hopes for
        Better Times," *Wall Street Journal*, June 14, 1985, p. 6.

44   In his younger days: "When You Say Busch, You've Said It All,"
        Business Week, Feb. 17, 1986, p. 58.

45   "The feeling of being conquered": Clausewitz, *On War*, p. 338.

46   "It might have been stupidity": "Superman Noha Ensures CNA
        Survival," Chicago *Tribune*, June 17, 1985.

48   "Nobody wants to wait": "Gerber Takes Risky Stance As Fears
        Spread about Glass in Baby Food," *Wall Street Journal*, March 6,
        1986.

54   One commentator has referred to him: Fuller, *A Military History
        of the Western World*, p. 184.

56   "he has an absolutely incredible ability": "Brassy Plan Suits
        Trump," *USA Today*, Nov. 20, 1985.

56   Arthur Rock: "Backer of Dreams," *Wall Street Journal*, Dec. 31,
        1985, p. 8.

57   In his study: Richard E. Neustadt, *Presidential Power* (New York:
        John Wiley & Sons, 1964), p. 147.

58   His dying words were: Basil Liddell Hart, *The Sword and the Pen*
        (New York: Thomas Y. Crowell, 1976), p. 183.

58   His contemporary Caulaincourt wrote: Quoted in Fuller, *A Military History of the Western World*, p. 411.

CHAPTER THREE

Maintaining Your Objective; Adjusting Your Plan

63   "Strategy concepts need to be explicit": Bruce D. Henderson, *Henderson on Corporate Strategy* (New York: New American Library, 1982), p. 6.

64   For example, during the Russian campaigns: F. W. von Mellenthin, *German Generals of World War II* (Norman, Okla.: University of Oklahoma Press, 1977), p. 284.

64   In the middle nineteenth century: Maury Klein, *The Life and Legend of Jay Gould* (Baltimore: Johns Hopkins University Press, 1986).

65   The authors of the book: C. Roland Christensen, Kenneth R. Andrews, and Joseph L. Bower, *Business Policy: Text and Cases* (Homewood, Ill.: Richard D. Irwin, 1978), p. 761.

67   In their book: Richard Tanner Pascale and Anthony G. Athos, *The Art of Japanese Management* (New York: Simon & Schuster, 1981), pp. 177–99.

67   "You don't just join": "Family Feeling at Delta Creates Loyal Workers, Enmity of Unions," *Wall Street Journal*, July 17, 1980, p. 13.

71   Q. T. Wiles: "Q. T. Wiles Revives Sick High-Tech Firms with Strong Medicine," *Wall Street Journal*, June 23, 1986, p. 1.

75   "much more strength of will": Clausewitz, *On War*, p. 243.

76   "Even competent generals": John R. Elting, *The Superstrategists* (New York: Charles Scribner's Sons, 1985), p. 1.

78   "change its way of doing business and thinking": "Mercury Hits New Heights in Old Surroundings," Chicago *Tribune*, July 7, 1986.

80   For example, when General Foods: Berg, *Mismarketing*, p. 67.

80   "We're not here to make profits": "A Troubled Polaroid Is Tearing
        Down the House That Land Built," *Business Week*, April 29,
        1985, p. 51.

85   "I do not favor pitched battles": Maurice de Saxe, *My Reveries on
        the Art of War* from *Roots of Strategy*, ed. Thomas R. Phillips
        (Harrisburg, Pa.: Military Service Publishing Company, 1940),
        p. 298.

88   At Austerlitz (1805), Napoleon's plan: Peter Young, ed., *Great
        Generals and Their Battles* (New York: Military Press, 1984),
        p. 16.

91   "A company's competitive situation": "Corporate Odd Couples,"
        Business Week, July 21, 1986, p. 100.

95   Studies show that companies that plan: Alan C. Filley and Robert
        J. House, *Managerial Process and Organizational Behavior* (Glen-
        view, Ill.: Scott, Foresman & Co., 1969), p. 206.

96   It was always best: A. L. Sadler, *The Maker of Modern Japan*
        (Rutland, Vt.: Charles E. Tuttle, 1937), p. 279.

97   According to Jack Welch: Robert H. Hayes, "Strategic Planning
        —Forward in Reverse?" *Harvard Business Review*, Nov.–Dec.
        1985, p. 114.

102   "One conclusion that has appeared": Henry Mintzberg and James
        A. Waters, "Researching the Formation of Strategies: The History
        of Canadian Lady, 1939–1976," in *Competitive Strategic Manage-
        ment*, ed. Robert Boyden Lamb (Englewood Cliffs, N.J.: Prentice-
        Hall, 1984), p. 90.

102   "Departures from tradition": Rosabeth Moss Kanter, *The Change
        Masters* (New York: Simon & Schuster, 1983), p. 291.

103   states Walter K. Joelson: "The Major Appliance Industry Is on
        Fast Cycle," *Business Week*, Sept. 2, 1985, p. 58.

106   "There seems to be a widespread assumption": John D. C. Roach,
        "From Strategic Planning to Strategic Performance: Closing the
        Achievement Gap," in Lamb, ed., *Competitive Strategic Manage-
        ment*, p. 219.

106   "It is evident": Clausewitz, *On War*, p. 306.

107   "You have a plan": S. L. A. Marshall, *Men Against Fire* (New York: William Morrow, 1947), pp. 196–97.

108   The American wine industry: "Tough Times: Wine Industry Finds It May Be Its Own Worst Enemy," *Wall Street Journal*, Mar. 19, 1986.

CHAPTER FOUR

Concentrating Greater Strength at the Decisive Point

115   "it seems to me": cited in Stefan T. Possony and Etienne Mantoux, *Makers of Modern Strategy*, ed. Edward Mead Earle (Princeton, N.J.: Princeton University Press, 1971), p. 229.

115   On a spring day in 1921: For background see John McDonald, *The Games of Business*, pp. 32–69.

120   According to Beech-Nut: "Baby-Food Fight: Beech-Nut, Heinz Put the Heat on Gerber," *Wall Street Journal*, July 24, 1986.

123   "Selling focuses on the needs of the seller": Theodore Levitt, "Marketing Myopia," *Harvard Business Review*, July–Aug. 1960, pp. 45–56.

126   At stake in the competition: "Bar Wars: Hershey Bites Mars," *Fortune*, July 8, 1985, pp. 52–57.

127   A product may flourish here: "Different Folks, Different Strokes," *Fortune*, Sept. 16, 1985, pp. 65–68.

128   Brokerage house Edward D. Jones & Company: "To This Broker Paris Is in Illinois, New York in Limbo," *Wall Street Journal*, June 12, 1986, p. 1.

129   "The opportunities on the blade side": "Gillette Keys Sales to Third World Tastes," *Wall Street Journal*, Jan. 23, 1986.

130   Kenneth Ohmae: "Rethinking Global Corporate Strategy," *Wall Street Journal*, April 24, 1985, p. 18.

134   "We are a very focused company": "State Farm Makes It a Policy to Stay No. 1," Chicago *Tribune*, Sept. 30, 1985.

140   Darwin Smith: *Wall Street Journal*, May 1, 1985.

143   Says Robert Barney: "Smaller Fast-Food Chains Brace for Shake-out," *Wall Street Journal*, May 31, 1985, p. 6.

145   In one study: T. N. Dupuy, *Numbers, Predictions and War* (New York: Bobbs-Merrill, 1979), p. 15.

151   Said David F. Miller: Chicago *Sun-Times*, April 28, 1986, p. 54.

152   "These companies have strategies": Richard P. Rumelt, *Strategy, Structure and Economic Performance* (Cambridge, Mass.: Graduate School of Business Administration, Harvard University, 1974), p. 123.

CHAPTER FIVE

Taking the Offensive and Maintaining Mobility

160   Military research has shown: Dupuy, *Numbers, Predictions and War*, p. 116.

161   John Smale: Priscilla Hayes Petty, "Behind the Brands at P&G," *Harvard Business Review*, Nov.–Dec. 1985, p. 89.

167   "We made the decision": "Cummins Cut Headed Off the Japanese," Chicago *Tribune*, March 3, 1986.

169   "No one can out-Big": "Firm Tries to Be David to Big Eight's Goliath," Chicago *Sun-Times*, Feb. 27, 1985, p. 55.

170   "People knew for years": "Want to Wake Up a Tired Old Product? Repackage It," *Business Week*, July 15, 1985, p. 130.

177   Hostile columns of infantry: Sir William Napier, cited in Burne, *Art of War on Land*, p. 139.

178   "The first system": "How to Keep Customers Happy Captives," *Fortune*, Sept. 2, 1985, p. 42.

183   Throughout history this truth: David J. Rogers, *Fighting to Win* (New York: Doubleday, 1984), p. 45.

183   "If we cast a glance": Clausewitz, *On War*, p. 291.

190   one study revealed: Leo Bogart, *Strategy in Advertising* (New York: Harcourt, Brace & World, 1967), p. 75.

191   During the Civil War: S. Dumas and Vedel Peterson, *Losses of Life Caused by War* (Oxford, England: Clarendon Press, 1923), p. 45.

191   the Allies suffered about double: L. P. Ayres, *The War with Germany* (Washington, D.C.: U.S. Government Printing Office, 1919), p. 119.

193   "I'll never forget the look on his face": "When You Say Busch, You've Said It All," *Business Week*, Feb. 17, 1986, p. 63.

194   "If you have a marketing department": "McDonald's Blasts Fast-Food Foes," Chicago *Tribune*, April 10, 1986.

196   Robert Galvin: Chicago *Sun-Times*, Aug. 1, 1985.

199   "The Coors pattern": "Why Chicago Is so Crucial to Coors," Chicago *Tribune*, April 7, 1985.

CHAPTER SIX

Following the Course of Least Resistance

201   "Normal soldiers": B. H. Liddell Hart, *Strategy* (New York: New American Library, 1974), pp. 25.

203   "The history of strategy": *Ibid.*, p. xix.

203   "The object is to defeat the enemy": Brian Bond, *Liddell Hart* (New Brunswick, N.J.: Rutgers University Press, 1977), p. 56.

204   "Marketing failures": Berg, *Mismarketing*, p. 220.

204   "Some companies seem to view": Michael E. Porter, *Competitive Strategy* (New York: The Free Press, 1980), p. 82.

204   in the three hundred campaigns: John Laffin, *Links of Leadership* (New York: Abelard-Schuman, 1970), p. 31.

211   "The growth of large industrial enterprises": S. J. Prais, *The Evolution of Giant Firms in Britain* (New York: Cambridge University Press, 1981), p. 1.

212   "factors systematically favouring": *Ibid.*, p. 40.

219   "The big boys in the industry": "Alberto's a Thing of Beauty at 30," Chicago *Tribune*, Feb. 24, 1986.

228   "We have achieved extraordinary results": "Success Just Keeps Rolling at Print Firm," Chicago *Tribune*, Feb. 18, 1986.

CHAPTER SEVEN
Achieving Security

232   Arthur Nielsen, Jr.: Berg, *Mismarketing*, p. 11.

236   "framework for competitor analysis": Porter, *Competitive Strategy*, pp. 47–65.

242   "He had copies of every speech": "Irv Hockaday Is Leaving His Mark on Hallmark: Diversity," *Business Week*, Nov. 12, 1984, p. 76.

243   "Our strategy sessions": McDonald, *The Game of Business*, pp. 158–59.

244   "victory, if achieved": Henderson, *Henderson on Corporate Strategy*, p. 28.

244   "penetrate his opponent's brain": Laffin, *Links of Leadership*, p. 28.

245   On Friday, February 21: "Eastern Air's Burman Badly Underestimated Obduracy of Old Foe," *Wall Street Journal*, Feb. 25, 1986, p. 1.

246   "We must acknowledge": Christensen, Andrews, and Bower, *Business Policy*, p. 408.

247   "I'm a doer": "Movers and Shapers," *Wall Street Journal*, Feb. 24, 1986, p. 148.

251   A recent issue: "Industrial Espionage," *Small Business Report*, Sept. 1986, pp. 44–47.

253   "We didn't realize": "Tupperware Aims at Keeping Abreast of Modern Times," *Wall Street Journal*, May 3, 1985, p. 2.

258   "They (Hasbro) do little or no research": "How Hasbro Became King of the Toymakers," *Business Week*, Sept. 22, 1986, p. 91.

259   George Mahrlig: "High-Tech Shocks in Ad Research," *Fortune*, July 7, 1986, p. 59.

262   "I don't think consumers": "Makers Test Concept of Autos With Adaptable Personalities," *Wall Street Journal*, Feb. 21, 1986, p. 21.

265   "We believe our strength": "Meanwhile P&G Is Losing One of Its Best Customers," *Wall Street Journal*, May 1, 1985, p. 31.

271   "a deterioration of mental efficiency": Irving L. Janis, *Victims of Groupthink* (Boston: Houghton Mifflin, 1972), p. 9.

271   In the book *Victims of Groupthink*: *Ibid.*, p. 10.

274   "bursting like a whirlwind": Stefan T. Possony and Etienne Mantoux, "Du Picq and Foch: The French School," in *Makers of Modern Strategy*, ed. Edward Mead Earle (Princeton, N.J.: Princeton University Press, 1971), p. 229.

280   "There are many fine generals": cited in Possony and Mantoux, *Makers of Modern Strategy*, p. 224.

281   "When troops once realize": Colonel G. F. R. Henderson, cited in Marshall, *Men Against Fire*, p. 170.

281   "Soldiers who have ceased to hope": Marshall, *Men Against Fire*, p. 170.

CHAPTER EIGHT

Making Certain All Personnel Play Their Part

284   "an army should always": Liddell Hart, *Strategy*, p. 328.

288   Noha: "Superman Noha Ensures CNA Survival," Chicago *Tribune*, June 17, 1985.

289   In her book: Kanter, *The Change Masters*, p. 41.

291   "As a general rule": B. Charles Ames, "Manager's Journal," *Wall Street Journal*, Jan. 13, 1986.

292   In his landmark book: Alfred D. Chandler, *Strategy and Structure* (Cambridge, Mass.: M.I.T. Press, 1962).

292   In their famous study: Paul R. Lawrence and Jay W. Lorsch, *Organization and Environment* (Homewood, Ill.: Richard D. Irwin, 1969).

295   "surprisingly, organization design": Boris Yavitz and William H. Newman, *Strategy in Action* (New York: The Free Press, 1982), p. 142.

296   A popular misconception: Benjamin S. Bloom, ed., *Developing Talent in Young People* (New York: Ballantine, 1985).

300   "there is in strategy": Clausewitz, *On War*, p. 283.

304   During the second battle: Burne, *Art of War on Land*, p. 11.

304   Research on combat: Marshall, *Men Against Fire*, p. 60.

305   in a *Harvard Business Review* article: J. Sterling Livingston, "Myth of the Well-Educated Manager," *Harvard Business Review*, Jan.–Feb. 1971.

305   Edwin H. Land: Quoted in A. D. Moore, *Invention, Discovery and Creativity* (New York: Doubleday, 1969), p. 36.

312   John Smale: Priscilla Hayes Petty, "Behind the Brands at P&G," *Harvard Business Review*, Nov.–Dec. 1985, p. 82.

312   One research study: H. G. Kaufman, "Relationship of Early Work Challenge to Job Performance, Professional Contributions, and Competence of Engineers," *Journal of Applied Psychology*, June 1974, pp. 377–79.

314   "It's essential that we take more risks": "Kodak Is Trying to Break Out of Its Shell," *Business Week*, June 10, 1985, p. 92.

315   "Four brave men": Ardant du Picq, *Battle Studies*, trans. John N. Greely and Robert C. Cotton (Harrisburg, Pa.: Stackpole, 1958), p. 110.

INDEX

329

DAVID J. ROGERS is president of Service
Innovations Corporation, a consulting firm
based in Highland Park, Illinois. For more
than fifteen years he has conducted seminars
in sales marketing, personal motivation,
and human-resource development. He is
the author of *Fighting to Win*.